THE GEOGRAPHY OF RURAL CHANGE

THE GEOGRAPHY OF RURAL CHANGE

Edited by
BRIAN ILBERY

An imprint of **Pearson Education**

Harlow, England · London · New York · Reading, Massachusetts · San Francisco · Toronto · Don Mills, Ontario · Sydney
Tokyo · Singapore · Hong Kong · Seoul · Taipei · Cape Town · Madrid · Mexico City · Amsterdam · Munich · Paris · Milan

Pearson Education Limited
Edinburgh Gate
Harlow
Essex CM20 2JE
England

and Associated Companies throughout the world

Visit us on the World Wide Web at:
http://www.pearsoneduc.com

First published 1998

ISBN 0 582 27724 8

British Library Cataloguing-in-Publication Data

A catalogue record for this book is available from the British Library

Library of Congress Cataloging-in-Publication Data

A catalog record for this book is available from the Library of Congress

10 9 8 7 6 5 4 3
05 04 03 02 01

Set by 35 in 10/12 pt Times
Printed in Malaysia, VVP

CONTENTS

FIGURES

TABLES

PREFACE

The origins of this book lie in a conference on the processes of rural change organized by the editor on behalf of the Rural Geography Study Group of the Royal Geographical Society with the Institute of British Geographers. Three years later a textbook has been produced on the often complex processes and outcomes of change experienced by rural areas in developed market economies. The focus is on both the multidimensional nature of rural change and the spatial manifestations of the recent period of major restructuring in the post-productivist countryside. Agriculture is continuing to experience a relative decline in importance in comparison with other, often consumptive, uses of rural space. Rural areas have thus become zones of land-use competition and potential arenas of conflict and tension.

This book does not adopt either an all-encompassing definition of *rurality* or one particular theoretical framework. Instead, contributors have been selected deliberately to emphasize the diversity of approaches to studying the geography of rural change. Relevant concepts, principles and theories of rural change are developed where appropriate, with case-study evidence drawn from different parts of the developed world. Above all else, the book aims to provide up-to-date insights into the economic, social, political and environmental factors affecting rural restructuring.

It is hoped that the book will be useful to both undergraduate and postgraduate students of geography and other social sciences. Most people with an interest in rural areas should find items of relevance. I gratefully acknowledge the helpful comments made by referees on an earlier draft of the text and I would like to thank Addison Wesley Longman for their patience and understanding.

April 1997   Brian Ilbery

ACKNOWLEDGEMENTS

We are grateful to the following for permission to reproduce copyright material:

The Geographical Association for figure 4.3; The Revue Forestière Française for figure 6.1; Academic Press Inc. for figure 7.2; Elsevier Science Ltd. for figures 7.3, 9.2 and 9.3; Routledge for figure 9.4; and John Wiley & Sons Ltd. for table 9.1.

Whilst every effort has been made to trace copyright holders, in a few cases this has proved impossible and we would like to take this opportunity to apologise to any copyright holders whose rights we may have unwittingly infringed.

LIST OF CONTRIBUTORS

Ian Bowler (Chapter 4) is Reader in Human Geography at the Department of Geography, University of Leicester. His research interests include the sustainability of rural systems, agricultural change under the Common Agricultural Policy, and regional images and the marketing of quality products.

Richard Butler (Chapter 10) is the recently appointed Professor of Tourism at the School of Management Studies, University of Surrey (previously at the University of Western Ontario). His primary research interests include the development of tourism and associated impacts, seasonality in tourism supply and demand, and the dynamics of tourism, with particular reference to insular and remote areas.

Owen Furuseth (Chapter 11) is Professor of Geography at the Department of Geography and Earth Sciences, University of North Carolina at Charlotte. His research interests are focused on community development issues and strategies, with a particular concern for rural economic and social planning, rural sustainability, and rural landscape change.

Andrew Gilg (Chapter 9) is Reader in Countryside Planning at the Department of Geography, University of Exeter. His research interests are in land-use planning, development control and housebuilding in the countryside. He is currently conducting research into the way in which farming has evolved and its effect on rural landscapes and environment.

Brian Ilbery (Chapters 1, 4, 12) is Professor of Human Geography at the Department of Geography, Coventry University. His research interests are in aspects of agricultural change and policy, notably farm diversification and land diversion. He is currently coordinating an international research project on regional images and the promotion of quality products and services in the lagging regions of the European Union.

Gareth Lewis (Chapter 7) is Professor of Human Geography at the Department of Geography, University of Leicester. His research interests are in social and behavioural geography, with particular interest in migration and household change in the countryside.

Terry Marsden (Chapter 2) is Professor of Environmental Policy and Planning at the Department of City and Regional Planning, University of Wales, Cardiff. His research interests are in rural restructuring and planning, and international aspects of agricultural and sustainable development. He is currently completing projects on food regulation and retailing in the United Kingdom.

Alexander Mather (Chapter 6) is Professor of Geography at the Department of Geography, University of Aberdeen. His research interests are in rural land use and environmental management, with particular reference to forest-related issues in Scotland, the United Kingdom and globally.

David North (Chapter 8) is Reader in Local Economy at the School of Geography and Environmental Management, Middlesex University. His research interests are in the field of regional and local economic development. He is currently working on the contribution of small businesses to local economic regeneration, especially the role of innovation and technology in rural SMEs.

Martin Phillips (Chapter 3) is Lecturer in Human Geography at the Department of Geography, University of Leicester. He has research interests in rural social and cultural geography, with a particular emphasis on rural gentrification, the rural middle class, and media images of the countryside.

Clive Potter (Chapter 5) is Lecturer in Environmental Management and Policy in the Environment Department at Wye College, University of London. His research interests are on the environmental impacts of modern agriculture and the greening of agricultural policy in both the European Union and the United States. He is currently investigating the likely environmental consequences of agricultural liberalization in western Europe.

CHAPTER 1

DIMENSIONS OF RURAL CHANGE

Brian Ilbery

INTRODUCTORY SETTING

Rural change is multidimensional and the countryside in developed market economies can no longer be viewed as being on the margins of economic, social and political change. Indeed, rural areas are now at the centre of interest and debate (Rogers, 1993) and many of the processes of change stem from broader and more general socio-economic and political processes. As a consequence, policy makers are having to re-evaluate policies relating to 'rural space'. The countryside is increasingly an area of consumption as well as production and the switch away from a productivist philosophy means that farmers and other primary producers are looking for new ways of generating income. Alternative uses of rural space are developing, and as Cloke and Milbourne (1992, p. 360) comment, 'there is no longer one single rural space, but rather a multiplicity of social spaces that overlap the same geographical area'.

This book is concerned with detailing and understanding the often complex processes and outcomes of change which have recently been experienced by rural areas in developed market economies. It emphasizes the spatial manifestations of rural change and provides up-to-date insights into different dimensions of rural restructuring, written by experts in their respective fields. Different theories, concepts and principles are outlined where relevant, but the book deliberately avoids adopting any one particular and overriding theoretical framework. Indeed, with the exception of the two theoretical chapters (Chapters 2 and 3), authors have been encouraged to develop their chapters in whatever they consider to be the most appropriate manner. Students need to be aware of this diversity of approach, which reflects the nature of debate in rural geography. The result is a book in which some chapters provide essentially empirical reviews of particular dimensions of rural change, whereas others develop a stronger theoretical component. In keeping with the general tenor of this diversity of approach, the book does not adopt any standard definition of either *rural* or *rurality*. Rural space is more readily identifiable as non-urban space and, for the purposes of this book, encompasses any phenomena that the authors take to be rural. As a consequence, some chapters do not enter into any debate over what constitutes rural space (but see below).

Although the book presents an international perspective on rural change, it is deliberately focused on developed market economies. Any wider geographical coverage would make the material too general (often inappropriate) and the choice of examples too difficult. As it is, the text examines a number of broad themes which are applied to selected countries, especially the United Kingdom and the United States. This choice again reflects the expertise of the authors, but it does help to focus on countries that are among the most advanced in the transition towards post-productivism.

The rest of this brief introductory chapter is structured into three parts. The first is concerned with different concepts of rurality and rural space; it emphasizes the futility of attempting to provide an all-encompassing definition of *rural*. The second provides some initial insights into the main dimensions and processes of rural change, with their spatially uneven outcomes. Finally, the structure of the book is outlined and justified, with brief insights provided into the nature and content of the individual chapters.

RURALITY AND THE RURAL IDYLL

Defining *rural* and *rurality* has occupied much space in geographical writing, but there remains little chance of reaching an agreement on what is meant by the term *rural*. Over 10 years ago, Newby (1986) suggested that what constitutes rurality is wholly a matter of convenience. More recently, Pratt (1996, p. 71) remarked, 'there are many rurals' and 'a multiplicity of meanings of the term rurality'. Indeed, *rural* is a chaotic concept which is contested in terms of identifying the defining parameters of rural space (Hoggart, 1990; Halfacree, 1993; Pierce, 1996). Places, spaces, areas, people and lifestyles are just some of the terms which researchers have related to the rural.

Traditionally, rural spaces were defined in terms of what were seen as distinctive rural functions (Cloke and Milbourne, 1992). However, as both Shucksmith (1994) and Pierce (1996) remark, this represents just one end of a spectrum of definitions, with the other (more recent) end seeing *rural* as a social representation of reality, or a mental construct. The mental construct focuses on how the rural is experienced by those individuals who integrate visions of rurality into their everyday lives (Hoggart *et al.*, 1995). In a more detailed review, Halfacree (1993) identified four approaches to defining rural:

1. *Descriptive* assumes the existence of rurality and empirically describes, through the use of different parameters and measures, its socio-spatial characteristics. The index of rurality for England and Wales (Cloke, 1977; Cloke and Edwards, 1986) was developed through the statistical manipulation of census variables.
2. *Sociocultural* assumes that (low) population density in some way affects behaviour and attitudes. Again, the approach avoids defining *rural* itself and instead draws associations between social and spatial attributes. Pahl (1966) was among the first to indicate there was no simple dichotomy between urban

and rural; indeed, there are (increasingly) urban aspects of rural society which distort any clear relationship between place and society.

3. *Rural as a locality* assumes a distinctive type of locality. If rural localities are to be studied in their own right, they must be carefully defined according to those characteristics which make them rural. However, it is clear that rural places are not all the same, and Halfacree demonstrates that the ruralization of industry is the movement to low wage areas (which could be urban or rural) rather than a movement to rural areas *per se*. Thus the term *rural* itself lacks explanatory power; Hoggart (1990) described it as a chaotic conception and one that should be used with great care.

4. *Rural as social representation* relates to lay discourses of rurality and 'the words and concepts understood and used by people in everyday talk' (Halfacree, 1993, p. 29; Jones, 1995). In this approach, attention turns to how the rural is perceived; it is a social construct because the emphasis is placed on how the occupants of rural spaces construct themselves. This is becoming more important as the consumption function of rural areas increases, and Pierce (1996) sees the social construction of rurality as having a major impact on research questions, policy processes and the sustainability of the rural environment. In turn, Cloke and Milbourne (1992) emphasized the importance of scale in social constructions of rurality. At the national scale, rural areas are portrayed as traditional and a refuge from modernity, whereas regional images become commodified by television programmes. However, it is at the local scale where the dominant meanings of *rural* are negotiated and where the differences between individuals and groups are highlighted (Little and Austin, 1996).

Halfacree (1993, p. 34) concludes his review by suggesting that the 'quest for any single, all-embracing definition of the rural is neither desirable or feasible'. Nevertheless, *rural* remains an important category because behaviour and decision making are influenced by people's perceptions of rural (Halfacree, 1995). Indeed, Phillips and Williams (1984) justify the use of the term *rural* to counterbalance the predominance of urban social studies and because it is a category of analytical convenience. More fundamentally, rural areas do have particular features which give them a distinctive social character; these would include relatively low population densities, open country and extensive land uses, lack of access to major urban centres, loose networks of infrastructure, and relatively low numbers of workers in secondary and tertiary industries (Clout, 1993). However, this does not mean that all rural areas are the same and that country dwellers are a uniform group.

Closely linked with social representations of rurality is the concept of the rural idyll. This portrays a positive image surrounding many aspects of rural lifestyle, community and landscape. For example, the countryside represents an ideal society which is orderly, harmonious, healthy, secure, peaceful and a refuge from modernity. It is characterized by mutual cooperation and support, self-help and voluntary commitment (Rogers, 1993). The essence of the rural idyll is captured nicely in the following quotation from Little and Austin (1996, p. 102): 'Rural life is associated with an uncomplicated, innocent, more genuine society in which traditional values persist and lives are more real. Pastimes, friendships, family relations and even employment are seen as somehow more honest and authentic, unencumbered with the false and insincere trappings of city life or with their associated dubious values.' Such an image is purposely portrayed by the media and used as a marketing logo for a wide range of products.

However, as Little and Austin (1996) continue to emphasize, the rural idyll is created by the wealthy for the wealthy and reflects particular power relations in society. This increasingly relies on the notions of selectivity and exclusion which, in turn, have led policy makers to consistently overlook problems associated with rural life. Yet, both MacLaughlin (1987) and Cloke and Milbourne (1992) have demonstrated how certain groups living in the countryside experience considerable exploitation and deprivation, just as Philo (1992) has emphasized the need to consider the neglected rural 'others'. Not all people living in rural areas conform to the rural idyll of a white, heterosexual, middle-class male who is able and of sound mind; hence the post-modernist call for the study of difference and the neglected others (Murdoch and Pratt, 1993). Clearly, people living in rural areas behave in different ways, aspire to different goals and have unequal access to what they desire.

THE CHANGING NATURE OF RURAL AREAS

Rural areas are dynamic and constantly changing in response to a wide range of social, economic, environmental and political factors. The pace of change has quickened in recent years and rural areas in developed market economies are diversifying as a result of broader socio-economic transformations and societal modernization (Hoggart *et al.*, 1995). As a consequence, uneven development and increasing differentiation are now characteristic features of rural space. The details of the various dimensions and causes of rural diversification are provided in later chapters, but it is appropriate at this stage to provide a skeletal outline of what has been happening.

Economically, rural areas are no longer dominated in employment terms by farmers and landowners. Agriculture is being restructured and farmers are having to adjust to national and international processes of change which are reducing the importance of the previously dominant productivist ethos. This adjustment often takes the form of generating new sources of income from non-agricultural activities, either on or off the farm (Bateman and Ray, 1994; Ilbery *et al.*, 1996). However, it may also involve the relocalization of the agrofood system in which locally produced quality products and services, with real authenticity of geographical origin, are transferred to regional and national markets (Marsden, 1996). In this way, local socio-economic change becomes part of global processes of restructuring which, in this case, involve the manipulation of consumer demand. Indeed, rural areas are now important elements of international economic arenas and among the leading investment frontiers (Clout, 1993). Employment has been growing faster in rural areas than in urban areas, especially in manufacturing, hi-tech and service sectors and in terms of small and medium enterprises (SMEs) (North and Smallbone, 1993, 1996). These activities, together with such new uses of rural space as recreation, tourism, environmental conservation and retailing, are creating different power relationships and a range of development trajectories in the countryside (Murdoch and Marsden, 1994). Clearly, rural areas are now characterized by consumption as well as production activities.

Many of these economic transformations in rural areas are related to social changes associated with the in-migration of particular groups of people. As Murdoch and Marsden (1994, p. 231) note, 'processes of rural change and class formation are inextricably bound together'. The countryside has been repopulated, especially by middle-class groups, so the social make-up of rural areas is now 'disproportionately biased towards those who, in terms of their wealth, power and influence, are influential in deciding national policy and public opinion' (Rogers, 1993, p. 1). In particular, the service classes took advantage of relatively cheap housing in the 1960s and 1970s to colonize the countryside. This element of the middle class work in either the public sector or private economic and social services; they have relatively high incomes, high job security and high educational attainment (Cloke and Little, 1990).

Once installed, the service classes exert a strong influence over the social and physical nature of rural space. They dominate the housing market, pushing up prices and thus excluding many original families, who are driven into key settlements and urban areas. The villages become gentrified and the service classes gain increasing control over local development and protect what they perceive to be their rural idyll, which is usually not related to agriculture. Indeed, there may be conflict between the different fractions of the service class. Different social groups occupy distinct spaces in villages and other social groups who may wish to move into rural areas, such as New Age travellers and ethnic minorities, are not made welcome.

Mormont (1990) has highlighted the changing relationship between society and space in the countryside. The increasing mobility of people, goods and information has helped to erode local communities and open up the countryside to new uses. This in turn has led to the creation of new power relationships and 'actor networks' which are likely to be dominated by external rather than internal linkages (Munton, 1995; Murdoch and Marsden, 1995). Murdoch and Marsden (1994) have analysed these new uses in terms of key land development processes and, in a detailed study in Buckinghamshire, England, show how each land development process has its own logic and institutions, as well as its own network of actors, which is more likely to be either national or global than local. Many of these changes have coincided with a massive reduction in the influence of the state, especially the local state, upon rural lifestyles. Such deregulation has been accompanied by the privatization of many services and the growing importance of the adventitious middle classes in local politics and planning.

Social and economic changes in the countryside have brought increased pressures on rural resources and caused governments in many developed market economies to re-evaluate their policies for the countryside. Reregulation has become an important element in some areas, notably in relation to sustainability and environmental conservation (Lowe *et al.*, 1993). National and, increasingly, international initiatives have been introduced to protect rural landscapes and wildlife and to prevent water pollution and soil erosion.

Although the processes of change outlined briefly in this section create spatially uneven patterns of rural development, Murdoch and Marsden (1994) present four 'ideal types' which characterize the range of outcomes one might expect in the countryside:

1. *The preserved countryside* relates to accessible rural areas and is characterized by antidevelopment and preservationist attitudes. These concerns are expressed especially by the middle-class fractions, who will use their power through the local political system to preserve their rural idyll. In addition, demands made by the service classes will provide the basis for new developments in leisure, conservation and residential property.
2. *The contested countryside* lies outside the main commuter zones and farmers and development interests remain politically dominant, and hence are able to push through development proposals. However, these are increasingly opposed by incomers who adopt the same position as in the preserved countryside.
3. *The paternalistic countryside* where large private estates and farms dominate and where development is still controlled by established landowners. Diversification opportunities, needed to raise incomes, are implemented with relatively few problems. The landowners continue to take a long-term management view of their property and thus adopt a traditional paternalistic role.
4. *The clientelist countryside* characterizes remote rural areas where agriculture is dominant but dependent on state subsidy (e.g. less favoured area payments). Any external investment is likely to be dependent upon state aid, and local politics will be dominated by employment concerns and the welfare of the local community.

THE STRUCTURE AND CONTENT OF THE BOOK

Any book on the geography of rural change has to confront two initial considerations: the chosen thematic approach and the geographical focus. The two are related through the choice of authors. These are all from the English-speaking (developed) world, which helps to restrict the text to that part of the world and to give it continuity and a strong focus. In turn, this leads to the exclusion of themes which may be more relevant in other parts of the world. The book deliberately concentrates on economic and social aspects of change, at the expense of political and cultural dimensions; readers interested in politics and culture are referred to texts by Cloke and Little (1990, 1997).

The book is divided into three parts. Part 1 reviews theoretical approaches to rural restructuring from both economic and social perspectives. Many of the ideas developed here reappear in later chapters, although there is no attempt to present an overall conceptual framework for examining rural change. Part 2 contains three chapters on extensive uses of rural land, specifically agriculture, environmental conservation and forestry. All are linked, in part, through the use of agricultural land, although it is accepted that more could have been included, especially on agriculture. The final section analyses selected aspects of the changing rural economy and society. They include counterurbanization, industrialization, recreation and tourism, service provision and deprivation, and rural policy and planning. A short concluding chapter attempts to summarize the main findings and their implications for the future direction of rural change.

Chapters 2 and 3 are concerned with theoretical perspectives on the economic and social restructuring of rural areas. In Chapter 2, Terry Marsden examines recent economic restructuring trends in rural areas from a broad political and social economy perspective. After describing in more detail the four ideal types of countryside outlined earlier in this chapter, he emphasizes the importance of three themes: regulation, commoditization and actor spaces. Taking examples primarily from the United Kingdom, but also from the United States and western Europe where appropriate, a division is made between horizontal and vertical dimensions of rural restructuring. Horizontal dimensions are associated with the urban-to-rural shift in both people and non-agricultural service and industrial sectors, whereas vertical dimensions involve the restructuring of the international food system and the domination of food markets by powerful and increasingly concentrated upstream and downstream sectors.

In Chapter 3, Martin Phillips explores some of the social theories and philosophical arguments which have been circulating within rural social geography in recent years. The discussion focuses upon two major reconfigurations of thought since the early 1980s. The first involves the re-examination of such traditional themes as demographic change, counterurbanization and service provision through broadly Marxist forms of political economy analysis. In contrast, the second and more recent study of rural social life concerns the issues of social identity, social difference and the construction and reception of cultural images highlighted by post-modern and post-structural social theories.

The next three chapters deal with changes in the extensive use of rural land. In Chapter 4, Brian Ilbery and Ian Bowler examine the shift from agricultural productivism to post-productivism. Theoretical explanations and empirical evidence, drawn primarily from the European Union and the United States, are provided for both agricultural productivism and the transition to post-productivism. Different 'pathways of farm business development' are described, before set-aside and pluriactivity are given as examples of the post-productivist transition. Potential land use changes under a 'sustainable' agriculture are outlined at the end. Clive Potter continues the agricultural theme in Chapter 5 by examining its developing relationship with environmental conservation. The chapter compares the philosophical traditions and policy processes giving rise to various agri-environmental initiatives in the United Kingdom and the United States. It assesses the immediate and long-term implications for the future conservation of rural landscapes and questions the political sustainability of current countryside management programmes. Potter concludes by calling for a radical reassessment of what government support to agriculture is for and who deserves to receive it.

In Chapter 6, Alexander Mather considers two main themes associated with the changing role of forests. The first is concerned with the growth in forest area, especially the afforestation of surplus agricultural land. The second explores changing forest values and perceptions. People now look to forests for more than the production of wood, and demands have increased for such services as recreation and nature conservation. These shifts in values have been conceptualized in the term *post-industrial forest*, which recognizes there are demands for other forest goods and services apart from timber production. However, it is possible that the changing role and value of forests in developed market economies have been facilitated by the exploitation of forests elsewhere.

Chapters 7 to 11 examine selected dimensions of the changing rural economy and society. Gareth Lewis examines recent migration and social change in the countryside in Chapter 7. After a brief account of the main driving forces behind the rural turnaround, Lewis highlights the need to focus on the household and the changing composition of the population rather than just population growth as the vital component in the counterurbanization process. A behavioural perspective is required to understand why and how people move into and out of the countryside, although it is recognized this must be set within a wider context of constraints and restrictions. Life course analyses, including changes in the family life cycle, are advocated for gaining a deeper understanding of the migration process in the countryside.

Chapter 8, by David North, examines the evidence for the urban–rural shift in industry within developed market economies, especially in the United States and western Europe. Different explanations for rural industrialization are offered, before attention focuses on the contribution of both small businesses (SMEs) and large branch plants of multinational companies to the rural economy. Although accessible rural areas seem to have benefited most from rural industrialization, North believes the trend towards more flexible production systems and developments in information and communication technologies will reinforce the attractions for many business people of working and living in a rural setting.

Managing change in rural areas has become a key aim of planners and in Chapter 9 Andrew Gilg develops a model of policy planning and applies it in case studies of rural policy mechanisms in the United Kingdom and the United States. Rural policy is shown to be a creature of the culture in which it develops, with the preservationist attitude in the United Kindom contrasting with the lack of control and the underlying ideology of private property in the United States. The increasing ability to acquire and use the countryside for leisure and pleasure purposes represents one of the major elements of rural restructuring and is the focus of attention in Chapter 10. Robert Butler describes the growth in rural recreation and seeks explanations in terms of large-scale changes in society. One very important factor has been the deliberate utilization of rural areas in the entertainment media; the subsequent increase in the number of visitors is not always beneficial to local residents. The economic, environmental and sociocultural impacts on rural areas of a growth in recreation are outlined, before Butler highlights the relative paucity of policies for rural recreation and tourism. Yet tourism and recreation must be considered as major factors in the planning and development of rural areas in the twenty-first century.

The final theme to be examined, in Chapter 11, is service provision and social deprivation. Owen Furuseth assesses the impact of rural restructuring on public services, particularly housing and health care, as well as poverty and deprivation. Increasing social differentiation is one outcome of the restructuring process and Furuseth examines its effect in both economically distressed and prosperous rural areas. Different theories of rural deprivation are outlined, before case study evidence is provided on the rural poor, rural health services and rural housing. The greatest challenge for policy makers in the future is to construct rural development strategies that meet the human resource needs of rural communities and allow them to build a long-term sustainable future. Sustainable rural development is one of the themes developed in the brief concluding chapter, which examines the implications arising from the main findings in the book for the future direction of rural change.

REFERENCES

Bateman, D. and Ray, C. (1994) Farm pluriactivity and rural policy: some evidence from Wales. *Journal of Rural Studies*, **9**, 411–27.

Cloke, P. (1977) An index of rurality for England and Wales. *Regional Studies*, **11**, 31–46.

Cloke, P. and Edwards, G. (1986) Rurality in England and Wales: a replication of the 1971 index. *Regional Studies*, **20**, 289–306.

Cloke, P. and Little, J. (1990) *The rural state?* Oxford University Press, Oxford.

Cloke, P. and Little, J. (eds) (1997) *Contested countryside cultures*. Routledge, London.

Cloke, P. and Milbourne, P. (1992) Deprivation and lifestyles in rural Wales II: rurality and the cultural dimension. *Journal of Rural Studies*, **8**, 359–71.

Clout, H. (1993) *European experience of rural development*. Rural Development Commission, London.

Halfacree, K. (1993) Locality and social representation: space, discourse and alternative definitions of the rural. *Journal of Rural Studies*, **9**, 23–37.

Halfacree, K. (1995) Talking about rurality: social representations of the rural as expressed by residents of six English parishes. *Journal of Rural Studies*, **11**, 1–20.

Hoggart, K. (1990) Let's do away with rural. *Journal of Rural Studies*, **6**, 245–57.

Hoggart, K., Buller, H. and Black, R. (1995) *Rural Europe: identity and change*. Edward Arnold, London.

Ilbery, B., Healey, M., Higginbottom, J. and Noon, D. (1996) Agricultural adjustment and business diversification by farm households. *Geography*, **81**, 301–10.

Jones, O. (1995) Lay discourses of the rural: developments and implications for rural studies. *Journal of Rural Studies*, **11**, 35–50.

Little, J. and Austin, P. (1996) Women and the rural idyll. *Journal of Rural Studies*, **12**, 101–11.

Lowe, P., Murdoch, J., Marsden, T., Munton, R. and Flynn, A. (1993) Regulating the new rural spaces: the uneven development of land. *Journal of Rural Studies*, **9**, 205–22.

McLaughlin, B. (1987) Rural policy into the 1990s: self-help or self-deception? *Journal of Rural Studies*, **3**, 361–64.

Marsden, T. (1996) Rural geography trend report: the social and political bases of rural restructuring. *Progress in Human Geography*, **20**, 246–58.

Mormont, M. (1990) Who is rural? Or how to be rural: towards a sociology of the rural. In Marsden, T., Lowe, P. and Whatmore, S. (eds) *Rural restructuring*. David Fulton, London.

Munton, R. (1995) Regulating rural change: property rights, economy and environment – a case study from Cumbria, UK. *Journal of Rural Studies*, **11**, 269–84.

Murdoch, J. and Marsden, T. (1994) *Reconstituting rurality*. UCL Press, London.

Murdoch, J. and Marsden, T. (1995) The spatialization of politics: local and national actor-spaces in environmental conflict. *Transactions of the Institute of British Geographers*, **20**, 368–80.

Murdoch, J. and Pratt, A. (1993) Rural studies: modernism, postmodernism and the 'post-rural'. *Journal of Rural Studies*, **9**, 411–27.

Newby, H. (1986) Locality and rurality: the restructuring of rural social relations. *Regional Studies*, **20**, 209–15.

North, D. and Smallbone, D. (1993) *Small businesses in rural areas*. Rural Development Commission, London.

North, D. and Smallbone, D. (1996) Small business development in remote rural areas: the example of mature manufacturing firms in northern England. *Journal of Rural Studies*, **12**, 151–67.

Pahl, R. (1966) The rural – urban continuum. *Sociologia Ruralis*, **6**, 299–327.

Phillips, D. and Williams, A. (1984) *Rural Britain: a social geography*. Blackwell, Oxford.

Philo, C. (1992) Neglected rural geographies: a review. *Journal of Rural Studies*, **8**, 193–207.

Pierce, J. (1996) The conservation challenge in sustaining rural environments. *Journal of Rural Studies*, **12**, 215–29.

Pratt, A. (1996) Discourses of rurality: loose talk or social struggle? *Journal of Rural Studies*, **12**, 69–78.

Rogers, A. (1993) *English rural communities: assessment and prospect for the 1990s*. Rural Development Commission, London.

Shucksmith, M. (1994) Conceptualising post-industrial rurality. In Bryden, J.M. (ed) *Towards sustainable rural communities*. University of Guelph, Guelph, pp. 125–32.

PART 1

THEORETICAL APPROACHES TO RURAL RESTRUCTURING

CHAPTER 2

ECONOMIC PERSPECTIVES
Terry Marsden

INTRODUCTION: THE PRETEXT

The past two decades in Britain, and more generally in the advanced world, have been a period when it has become more difficult to take rural areas and their economic activities for granted. As in other branches of the national and international economy, the ideology of progress, and particularly of economies of scale, tended to be automatically accepted as a driving force which defined rural space. The post-war settlement between the twin goals of boosting agricultural output after the hungry experiences of the wartime U-boat campaign and the need to protect the countryside and contain urban ribbon development (formalized in the Agriculture Act and the Town and Country Planning Act, both of 1947) provided a backcloth for a period of agricultural expansionism which was to go largely unquestioned until the dawn of the 1980s. Many scholars have now documented this period well (Newby, 1979; Marsden *et al.*, 1993; Winter, 1996) and there is little need here to reiterate the generalities of development. Agricultural production and land use became the pole around which the 'merry-go-round' of other economic activities would swing.

And, while there was a growing demand for the protection of greater slices of rural space, through the further development of protected and amenity areas, as long as both Whitehall and Brussels were prepared to continue supporting farm prices, farmers – the central figures in the rural economy – saw it as their duty to continue to produce food products using mechanized inputs and a decreasing amount of farm labour.

During this period, perhaps not surprisingly, theoretical understanding of rural space was largely confined to the work of agricultural economists. Having often been employed in agricultural economics departments at the behest of the Ministry of Agriculture, and like the farmers themselves, the economists were under a national obligation to deliver analyses that would reduce the constraints on 'efficient' production. Commonly termed *agricultural adjustment*, the main rural economic problem was to restructure agriculture to follow economies of scale under quantitatively defined 'viability' parameters. Hence, many if not the majority of farms in the United Kingdom and in many other developed market economies were hooked

13

to a technological and economistic 'treadmill' (Cochrane, 1968; Dexter, 1977), which defined individual farms and rural areas as viable or subviable along varying continua. Thus rural and agricultural marginalization was defined around the productionist and efficiency criteria established by economists following a neoclassical economic paradigm. This was tied, somewhat ironically, to a highly regulated and state-supported policy sector.

This pretext is introduced here to help define the significance of theory to the development of contemporary rural space. During this period, human geographers were generally uninterested in 'theory'. Beyond making passing reference to Von Thünen and rural space as an isotropic plane (Chisholm, 1962), they preferred to leave such abstractions to the agricultural economist, and, by default, to a particular economic rationality. A more fruitful avenue was to follow a directly empirical route: knocking on the farmhouse door and arranging an interview with the farmer *after* he or she had already completed the farm business survey record and June census return. These were the basic raw materials for the teams of economists established strategically in the different regions of rural Britain. If this work did accommodate theory, it tended to begin to see the farmer in behavioural terms, stressing the 'non-economic' variables which were recognized as central to the real tapestry of rural and agricultural change at the time. This, however, was something of a lonely occupation, for it questioned the principles of economic rationality which, at the time, and despite increasingly heavy public support, had to dominate as a progressive rural ideology.

But during the succeeding decades, marked by the growing economic, environmental and political crisis surrounding agriculture (and more recently food), the role of theory for the rural economy has both shifted and been reconstructed by different groups of scholars less tied to, and sometimes highly critical of, the earlier agricultural economists' paradigm (Marsden *et al.*, 1986). This chapter will focus on some of the key theoretical advances from a broad political and social economy perspective during this more recent period; it will show the utility of theory for guiding research, policy and more effective interpretation of the contemporary rural economy. This is a period when new theoretical advances are needed and when human geographers learn to live with theory much more comfortably. Such developments, however, as in the postwar period, tend to reflect as well as guide the real changes that have recently occurred in British rural space. In particular, the rural 'merry-go-round' with agriculture at its hub looks decidedly shaky.

Moreover, rural geographers can no longer so neatly divide the chapters of their textbooks along the main land-using activities of the rural economy (e.g. agriculture, forestry, housing, conservation). The processes of economic restructuring which have taken hold in rural space over the past two decades do not necessarily abide by these categories of development, neither can they be effectively regulated by the national postwar policy compromise which prioritized agricultural production and urban containment. The period of progressive productivism has given way to what has been termed a 'post-productivist countryside'. This requires new theoretical understanding. The rest of this chapter will outline the main dimensions associated with recent rural economic change.

THEORIES FOR THE POST-PRODUCTIVIST COUNTRYSIDE

Since the early 1980s it has become increasingly clear that rural areas, both in the United Kingdom and in the advanced world, have been caught up in a much more complicated national and international political economy: a period of social and economic restructuring which has become highly diverse and fragmented (Marsden *et al.*, 1990; Marsden *et al.*, 1993). This has not meant that everything before it has been gradually swept away. For instance, the growth in the intensity of agricultural production, of technological advances such as genetic manipulation, and in the drift of agricultural labour from the land, have all continued in many parts of Britain and mainland Europe. Nevertheless, the central organizing frameworks established in postwar times, themselves based upon national strategies for urban as well as rural living (Hall *et al.*, 1972), have been largely overtaken by the tide of a rural (and urban) restructuring process which has been both economically and socially driven. Before looking at theoretical approaches to these processes, it is worth identifying some of the main features from a British perspective.

This can be done in general terms by approaching rural restructuring from both horizontal and vertical dimensions. Probably in the starkest of fashions, at the horizontal level there has been a radical urban-to-rural shift in both population and economic activity. These have worked together, with waves of ex-urban groups such as commuters, retirees and tourists unevenly creating demands for wider and more diffuse forms of rural service activity. In addition, the relocation or new formation of industry and services has in some areas taken the lead in attracting population growth and counterurbanization. The conventional forms of territorial planning have been unable to 'contain' these processes, as they have spilled over green belts and the established new towns, and socially and economically reconstituted many rural villages. This has redefined the relationships between town and country, placing rural space very much in its differential regional context; that is, as an essential part of regional territorial change.

Second, and this has been given considerable attention by scholars recently, the vast majority of rural land – because of its agricultural occupancy – is still hooked into a series of distant agricultural and food markets dominated increasingly by powerful and concentrated manufacturing and retailing sectors (Ward, 1990; Flynn and Marsden, 1992). In particular, the role of the 'downstream' sectors in shaping, at a distance, the rural land-based farm sector is a significant feature of recent agricultural changes. The current raft of policy measures emanating from Brussels and Whitehall, in terms of both production and environmental support, still focuses upon the land-based farmer as the principal recipient of public funding and the delivery of public goods in the countryside. Yet it is increasingly the non-agricultural actors which need attention in assessing land-based rural change.

The starting-point for theorizing about contemporary rural restructuring is to see rural change as the spatially variable condition which combines these vertical and horizontal dimensions. Indeed, much of the work over the past decade has focused upon how these processes are being organized in different localities, both by the combinations of actors *in situ* and by their sets of relationships with non-local actors and agencies. Such a perspective – broadly defined as new political economy – has begun to redefine the concept of rural space and rural restructuring itself. Rural space becomes a highly elastic phenomenon, constructed out of combinations and layers of social, political and economic relations, traversing different physical spaces at any one time (Mormont, 1990). It is differentially tied to the regional and international economy as much as to more traditional forms of national regulation (Lowe *et al.*, 1993).

To begin to theorize about contemporary rural change, therefore, requires a consideration of differential uneven development. This is only partly defined by the vagaries of agricultural change. It is just as associated with development processes which *combine* former economic sectors (e.g. housing, minerals, agriculture, amenity) in new and often novel ways. One example would be the development of farm conversion markets and more complex property uses associated with access agreements and farm tenure arrangements. During the period of agricultural modernization, rural areas were, to all intents and purposes, regarded by the majority of the urbanized population of countries like Britain, as marginal places or backwaters to progressive industrialism. Now, after at least 15 years of uncertainty in the policy arena, and 'boom and bust' economic cycles, perhaps a little ironically, some rural areas are seen as much more central to people's lives.

Food, the environment and the pressure for amenity are creating new and uneven demands on rural space; the accommodation of much broader consumption concerns, beyond those dealing simply with production, have begun to foster new types of rural and regional development. These new demands, for 'quality' food production, public amenity space, positional residential property, areas of environmental protection, and for the experience of different types of rural idyll or *urban antithesis* are now much more entrenched in rural space than a decade and a half ago. Moreover, they owe their origins to a disparate array of forces both within and beyond state control or guidance. This means that our theoretical approaches have to be adaptive to a realigned understanding of state–society relations, a situation whereby agricultural and planning policy now only provide part of the justification for development in the countryside. A key theoretical question, therefore, must be: what are the different spatial expressions of this new rural differentiation, and are there common factors which are actively making rural space different?

First to be outlined is a typology detailing this differentiation. Next comes an examination of some of the common dynamics characterizing this differential rural change. Although the reference point is the United Kingdom, there is emerging evidence of similar features developing both in parts of northern Europe and North America (Marsden, 1995; Marsden *et al.*, 1996). The theoretical developments outlined in the rest of this chapter thus raise important questions for international comparison as well as for differential rural change in the United Kingdom.

Differentiating rural change: emerging ideal types

Through theoretical development and ongoing empirical work, it is necessary for analytical purposes to consider the nature of rural differentiation through the development and refinement of typologies which attempt to capture the dynamic nature of rural change. One such typology is developed here for the purposes of situating the rest of the discussion and demonstrating the process of differentiation as an economically and socially active force in rural change.

A major theoretical and empirical challenge now faces scholars concerning the need to provide analyses of rural uneven development. One way to start this is to consider emerging *ideal types* – Weber (1949) termed them *Gedankenbild* – which characterize the range of expected outcomes from the key economic, social and political processes shaping the countryside. Such ideal types have to incorporate combinations of local, regional and distant relationships, and they have to consider how these relationships become established in space. In addition, they are associated with the social and economic *reaction to* as well as the *articulation of* economic change.

Four ideal types characterize the British countryside: the first, the *preserved countryside*, is perhaps most evident in the English lowlands, as well as in the attractive and increasingly accessible uplands. These areas are characterized by established preservationist and antidevelopment interests and local decision making. Although agriculture has recently been intensively practised, farmers are now seeing the benefits of diversification in serving the local demand from ex-urban groups. Concerns are most clearly expressed through middle-class fractions (Cloke and Thrift, 1990), who may impose their views through the planning system on would-be developers (Abram *et al.*, 1996). In addition, demand from these fractions provides the basis for new development activities (often of a very exclusive kind), associated with leisure, industry and residential property. Rural change is thus a highly contested process, articulated by different middle-class consumption interests who use the local political system to protect their environmental positional goods (Hirsch, 1978). These areas have also been subject to high levels of economic growth and development (Murdoch and Marsden, 1994) both in terms of inward investment and internal firm formation, through the development of new industrial units and the growth of a more diverse service sector as a consequence of the purchasing power of middle-class groups.

Second, the *contested countryside* is represented by those rural spaces which lie outside the core commuter catchments and, as yet, may be of no special environmental quality. Here farmers (as landowners) and development interests may still be politically dominant and are thus able to push through development proposals associated with agricultural diversification and small industrial schemes. These developments are increasingly opposed by the recent waves of incomers who adopt the positions that are effective in the preserved countryside. Thus, the development process is marked by increasing conflict between old and new groups (Ward *et al.*, 1995; Flynn and Lowe, 1994).

Third, the *paternalistic countryside* refers to areas where large private estates and farms still dominate and the development process is decisively shaped by established landowners and farmers. Many of the large estate owners and farmers may be facing falling farm incomes and are thus searching for new sources of income. In this sense many are trying to convert agriculturally productive capital into new forms of landed capital, by selling off redundant buildings or leasing land out to private farm management companies (Whatmore *et al.*, 1990). Landowners and farmers are unlikely to be constrained in these strategies and they still take a long and custodial view of their land, property and village contexts, adopting a modified stewardship role in the rural community. These areas are less likely to be under great development pressure than either of the earlier two types.

Fourth, the *clientelist countryside* is likely to be found in the remote upland rural areas where agriculture and its associated political institutions still hold sway, but where farming can be sustained only by state subsidy, such as less favoured areas' per capita payments and welfare transfers. Processes of rural development are dominated by farming, landowning, local capital and state agencies, usually working in close corporatist relationships. Farmers will depend on systems of direct agricultural and agri-environmental support, e.g. Environmentally Sensitive Area (ESA) payments, and any external investment is likely to be dominated by employment concerns and the welfare of the rural community.

The typology is an attempt to characterize the processes of uneven rural change as driven by different sets of internal and external powerful interests. These in turn tend to 'create their own spaces' by shaping rurality in different directions. In the paternalistic countryside one could expect landowners and farmers still to hold sway not only, as Newby *et al.* (1978) suggested, in demarcating and upholding cheap agricultural labour markets, but also in shaping the degree of agricultural diversification and the shift of assets between agricultural operations and other land development opportunities. Moreover, the coincidence of the MacSharry reforms of 1992 in creating greater incentives for set-aside and environmental conservation, coincided with the devaluation of UK national currency (and the green pound) and the relative uplift in European regulated price support levels. This provided a fillip for the intensive producer, further creating the conditions for land concentration, intensive production and contract farming in many parts of arable eastern England.

In addition, the patchy development of agri-environmental measures (such as farm stewardship and ESA schemes) in the intensively farmed areas has further recognized and reinforced the custodial property rights of farmers and landowners. Their traditional representative bodies (the National Farmers' Union and the Country Landowners' Association) are now more eager to assimilate the post-productivist countryside as long as it is shaped around the exploitation of their members' property assets (see below). In this sense, then, the significance of the landed classes in the paternalistic countryside has been far from a static phenomenon in the post-productivist era. Indeed, partly because of the incremental and partial nature of many of the policy reforms (i.e. failing to really address sustainability, and upholding agricultural freehold property rights), they have been able to maintain their local and regional grip in many of the most intensively farmed (and thus potentially unsustainable) bits of rural Britain. Agrarian paternalism has adapted to both local and external demands by accommodating the broader environmental stewardship of agricultural land. It is now reproduced around both an agricultural and environmental base. And, with the current absence of a rump of preservationist middle class, the degree to which they can truly deliver the alleged benefits of this broadened stewardship role remains largely unobserved at the local level.

In the clientelistic areas of upland marginal Britain, where rural economic development concerns sit side by side with growing formal environmental protection and continued agricultural state support as a multifaceted system of public welfare from Whitehall and Brussels, it is the growing dominance of the representatives of the 'provider' agencies which begins to shape developments. Also, these areas have traditionally been the sources of extractive industries (e.g. coal, water, energy), and with their physical remoteness and weak labour markets, they have been prone to the relocation of industries handling waste, toxics and defence contracts. The recent privatization of most of the utilities in Britain has also tended to increase reliance upon the private providers of jobs in these areas. As a result, in Wales and in parts of Cumbria, one now sees the emergence of some interesting juxtapositions of private and public clientelism, where national park authorities, mining firms and public and semiprivatized development agencies (such as English Partnerships), play influential roles in shaping developments. The relationships and dependencies between the provision of jobs and the provision of consumption spaces for amenity come into stark relief (Munton, 1995). And as a result of the weak labour markets and the dependencies they bring, in terms of the powers of inward investors to bargain with public authorities, it means the outcomes for environmental care and plan making are of a different order and shape than in the other types of countryside developed here.

Such emerging ideal types begin to capture the processes of differentiation occurring in the post-productivist countryside. They require further development and analysis. In considering such expressions of rural change, it is also necessary to explore the unifying theoretical concepts which help to provide a more coherent picture of this rural differentiation; that is, to begin to understand the causes of difference. The next section develops an outline of some key common dynamics needed to understand contemporary economic change in the rural sphere. Collectively, these conceptual parameters begin to shape an approach to the study of the rural economy, after the rural and agricultural crisis of confidence during the past 15 years. The following questions pervade this discussion: To what extent does this add up to a coherent picture of the contemporary rural economy? Does it suggest the demise of national rural structures and the differential emergence of regional ruralities? What sorts of concepts do we need when attempting to find answers?

THE SOCIAL AND POLITICAL ECONOMY OF COMTEMPORARY RURAL SPACE

The new centrality of rural space in regional and national economic terms derives at a general level from the changing bases in which economic relationships are *formed*. They are regulated by actors and agencies through social and cultural interactions, using different combinations of knowledge, artefacts, attributes, technologies and natural phenomena (Law, 1992). A host of regulationist and social constructivist literature in economics and economic geography has highlighted how market relations are socially and politically constructed and embedded in institutions, norms of behaviour and trust relationships (Grabher, 1993; Salais and Storper,

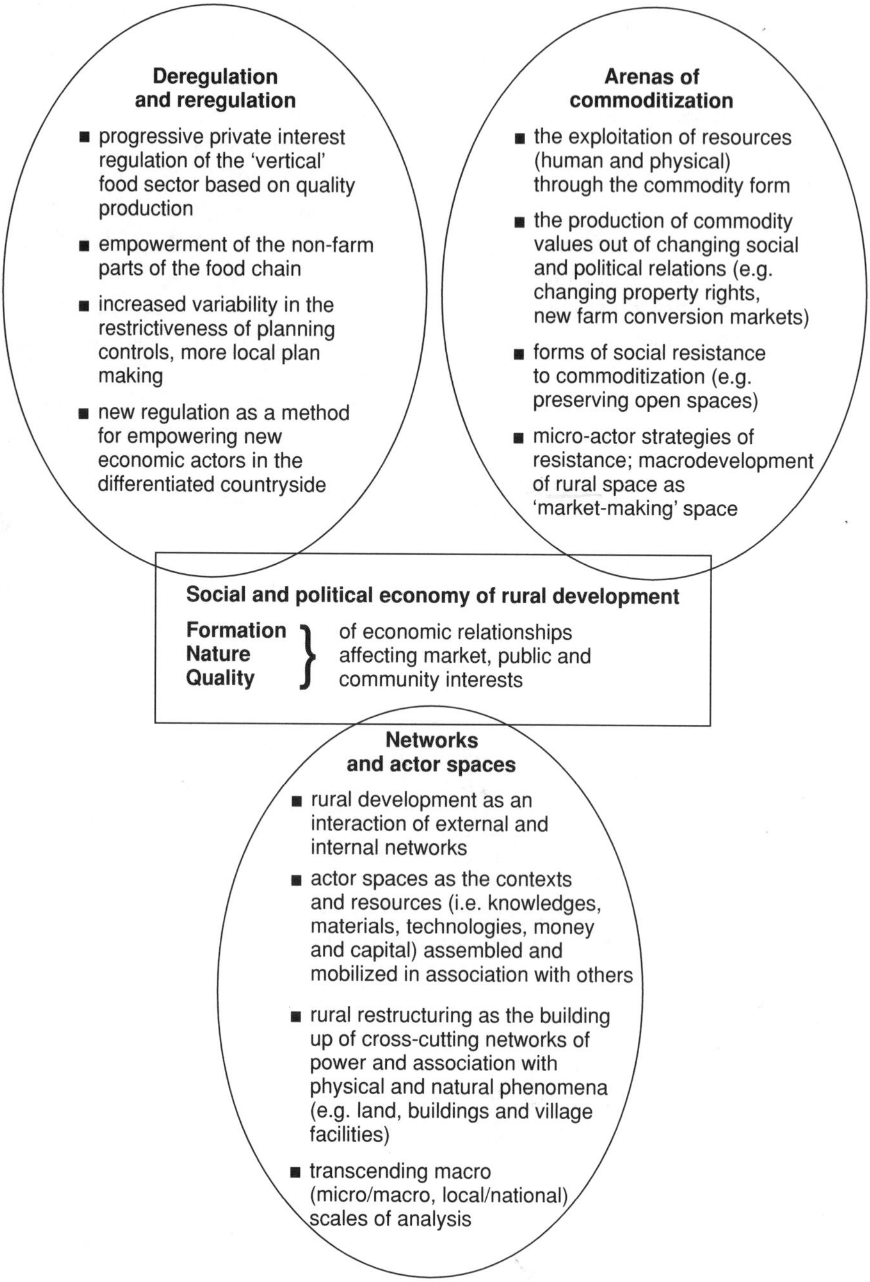

Deregulation and reregulation

- progressive private interest regulation of the 'vertical' food sector based on quality production

- empowerment of the non-farm parts of the food chain

- increased variability in the restrictiveness of planning controls, more local plan making

- new regulation as a method for empowering new economic actors in the differentiated countryside

Arenas of commoditization

- the exploitation of resources (human and physical) through the commodity form

- the production of commodity values out of changing social and political relations (e.g. changing property rights, new farm conversion markets)

- forms of social resistance to commoditization (e.g. preserving open spaces)

- micro-actor strategies of resistance; macrodevelopment of rural space as 'market-making' space

Social and political economy of rural development

Formation
Nature   } of economic relationships affecting market, public and
Quality   community interests

Networks and actor spaces

- rural development as an interaction of external and internal networks

- actor spaces as the contexts and resources (i.e. knowledges, materials, technologies, money and capital) assembled and mobilized in association with others

- rural restructuring as the building up of cross-cutting networks of power and association with physical and natural phenomena (e.g. land, buildings and village facilities)

- transcending macro (micro/macro, local/national) scales of analysis

Figure 2.1 Dimensions of rural development: social and political economy

1992; Amin and Thrift, 1994). Economic restructuring or uneven development thus has a social and cultural basis in the rural domain, both generally and in particular (Marsden *et al.*, 1996). We need to consider the nature and the quality of economic and power *relations* (Fig. 2.1).

This social economy perspective holds at least three components. First, it focuses not only on the patterns of uneven rural development but also on the ways in which combinations of market, public and community interests and networks carry forward the process of rural development. This is the social dynamic of rural restructuring. Second, it is also necessary to examine the ways in which these differentiating rural trajectories (such as those in the typology above) actively redefine combinations of local rural resources in new ways. As outlined below, this has been quite firmly associated with the *commoditization* dynamic, whereby rural resources have been attributed a market value for exchange. For instance, former agricultural resources (barns, housing, land, woods) are being unevenly exploited by new groups of actors; in addition, new rural amenity and niche products are being created through the reconstitution of place and identity. Third, a focus needs to be placed on the ways in which traditionally perceived economic relations become embedded and carried through different sets of social, political and regulatory actors and agencies. For instance, ESA payments and structures funding are regulated through the Ministry of Agriculture, Fisheries and Food (MAFF) down to farmers, and the Rural Development Commission in England allocates grant funding to particular areas over others. Increasingly, the institutional frameworks carrying economic benefits to rural areas are highly spatially variable.

The term *social economy* connotes the integration of these factors of dynamic rural change. This grounds the multiplicity of economic relations in their social and geographical setting (i.e methodological situationalism). It provides a theoretical basis for developing a systematic comparative approach to the new rural diversity. This deals with rural space as a holistic entity. Such an approach is particularly apposite for rural conditions, given the growing and diverse demands placed on rural resources and facilities and the degree to which they relate to different conceptions of quality for products, services and economic relationships. But this depends upon quality (often with specific reference to place of origin as an authentic locale) becoming a way of market entry and maintenance for many rural economic sectors. This relative success then influences overall development. Of course, not all areas benefit from these engagements in new exchange relations. Indeed, almost by definition, one can expect greater local and regional inequalities to be generated as less emphasis is placed upon a national system of social and economic provision. What are some of the key parameters in this social economy?

Deregulation and reregulation

The political economy of the 1980s and 1990s has been dominated by the attempts of government to deregulate market relations and to reduce state burdens by privatizing former state assets. This process has not slackened in the 1990s, with government attempting to reduce interference in economic activity and to restructure the institutions, including government departments and rural agencies. These processes have served to provide different competitive advantages and disadvantages to rural spaces. On the one hand, the agricultural corporatism which had dominated thinking on rural development over the postwar period has been undermined by the need to reduce financial subsidy, to eliminate marketing boards and wages boards and to encourage 'free trade'. In terms of the vertical food sector this has acted to empower the non-farm parts of the food chain (particularly retailers and processors). Through the development of new food safety legislation from Whitehall (e.g. the Food Safety Act 1990) and Brussels (e.g. the Hygiene of Foodstuffs Directive 1995), retailers have been awarded new powers to police the food sector by essentially private means (Marsden and Wrigley, 1996). The changes have led not so much to deregulation as to new forms of *private interest regulation*, whereby private sector actors and agencies take responsibility for the successful implementation of policy, ostensibly in the public interest but also in the private interest (Marsden and Wrigley, 1995).

On the other hand, in terms of the non-agricultural restructuring, early attempts to deregulate the planning system were vouchsafed by the rural middle classes keen to protect their increasingly positional grip on large chunks of Middle England. This subsequently led in the 1990s to the uneven development of local and regional power coalitions of interest which reinforced local plan making along conservationist lines. The Planning and Compensation Act 1991 determined that development control decisions should be made in accordance with the plan unless material conditions indicated otherwise, following the extension of local planning to all rural land under the Town and Country Planning Act 1990. Despite over a decade of neoconservative philosophy, the net effect has been for a more regulated countryside, not less regulated, and it is now subject to capture by particular local residential, producer and consumer interests within different spatial contexts.

More complex models of interest representation have developed over rural space, models which are sensitive to local social and economic conditions. For instance, agricultural changes in some regions have become more reliant upon the public-interest local planning system, such that farmers (in the buoyant growth regions like the South-East) are obliged to negotiate changes through the development control system. Passage through these systems of mediation is not just associated with political gains and losses. They moderate and regulate the flows of mobile and fixed capital into and out of rural space. They are carried by networks of actors who are differentially empowered by these regulatory changes. In many parts of the preserved countryside in south-east England, and increasingly in the contested spaces of the South-West, farmers become the bystanders to the powerful coalitions of middle-class fractions and non-agricultural entrepreneurs who have gained more than simply a physical foothold in rural space. Their power and legitimacy have grown in the face of both agricultural retrenchment and the onset of local planning systems which embrace middle-class protectionism in the face of an ideological questioning of planning regulation within the free-market orthodoxy.

Theoretically then, as with subsequent parameters, recent tendencies have been far from unilinear or cast simply and strictly along economistic lines. The new post-productivist countryside, a terrain where consumption and exchange vie seriously with the more traditional productive activities, becomes a contested social milieu, through which the mobility and fixity of capital occurs. The regulatory diversity, and some would argue fragmentation, has most recently been recognized with the publication of White Papers for England, Wales and Scotland. These White Papers explicitly admit that government cannot clearly and purposefully regulate the countryside in the traditional sense, even if regulation is now desired. Rather, as a result of the deregulation–reregulation dialectic of the past 15 years, all it can do is to give a voice to the sanctity of rural life and to rely upon the rural economy to prosper from its own resources, both physical and human.

Hence, the regulation–deregulation dynamic provides a very uneven playing field for economic activity in rural space. It changes the sources of competitive advantage and disadvantage at a locale's disposal. The different countrysides which constitute modern rural space are having to develop their own regulatory and institutional frameworks. For instance, a growing distinctiveness can be observed between the institutional frameworks operating between the different countrysides outlined in the ideal types above. This suggests the need to develop theoretical and conceptual constructs concerning what Clark (1992) has called 'real regulation' (Munton, 1995; Flynn and Marsden, 1995). It is necessary to understand more fully how regulatory activity and authority becomes *spatialized*, and how these spatializations begin to influence the flow, direction and density of economic relations.

Arenas of commoditization and social resistance

Although by no means the most inviting of terms, commoditization describes a concept that is central to the understanding of contemporary rural restructuring. The complex process of socially exploiting resources (particularly labour, but also increasingly land, nature and the built form of the countryside) through the extension of the marketized commodity form has its more recent origins in the sociology of agriculture literature. Here, following a conventional Marxist interpretation of the reasons for the maintenance of family farming and the progressive expansion of capitalist agriculture and agribusiness, the concept can now be more imaginatively applied to an actor-oriented perspective for rural restructuring. Commoditization represents a variety of social and political processes by which commodity values are constructed and attributed to, in this case, rural and agricultural objects, artefacts and people. It does not represent one all-encompassing process which, for instance, transforms agricultural labour processes or simply puts a price tag on a formerly publicly owned piece of property. Rather, it is important to see it as a diversely constructed phenomenon around which rural development processes coalesce then diffuse. It poses two questions about the complexity and diversity of contemporary rural space: How do commodity and other social values shape social practices? How are commodity relations and values generated, and challenged, through the active strategizing, network building and knowledge construction of particular producers, consumers and other relevant users of rural resources? It involves exploring how people, objects and combinations of natural and human artefacts are actively assembled in rural places, and how they are used and valued in market exchange. Conversely, it is also important to understand how some actions remain outside the commodity form. Many local action groups in rural areas assemble resources in order to fight these 'commoditization battles' with developers and planners. Murdoch and Marsden (1994) have documented at length how, in the preserved countryside of the South-East, these middle-class groups have held a strong influence in shaping the social and physical spaces in which they reside.

But below this generalized level of discussion concerning commoditization, it is necessary to examine in more specific detail how social actors assemble and construct 'value' in rural space, then to assess how much this seems to take on a commodity form. Explicit struggles may occur over the attribution of social meanings and values, access to resources and in relation to issues and differences of social identity. Hence to explore commoditization processes is to examine the means and strategies by which social resistance occurs around prevailing commodity definitions and practices. As with the discussion of regulation, it is by no means a predictable process. Rather, it highlights the socially active nature of what, at a distance, might be seen as economic relations. Hence, it is a central element in the changing social economy of rural space.

In some long-standing empirical work on rural areas in the United Kingdom (Marsden *et al.*, 1993; Murdoch and Marsden, 1994; Marsden *et al.*, 1996) the multidimensionality of this process for different rural spaces has begun to be documented. This provides a useful comparative device in understanding rural differentiation. The commoditization of redundant farm buildings, for example, (Kneale *et al.*, 1991) was highly spatially variable in the late 1980s. Similarly, the development of recreational sites, such as golf courses, tended to cluster in the preserved countryside of south-east England. This recommoditization of former agricultural land is differentially regulated by local planning systems such that high levels of (market-based) development pressure can be resisted in the preserved countryside, and albeit with considerable difficulty, they can be attracted to the poorer rural regions of the uplands and the North, through effective developmental planning. This demonstrates that developmental processes in the post-productivist countryside are far from simply reliant upon the uneven development of capital and economic activity. If this were the case, very different rural and economic landscapes would now be witnessed. Rather, people and their institutions have to 'carry' and resist the commoditization processes. And the balances between resistance and accommodation help to shape rural space both economically and physically.

More work needs to be conducted on the ways in which value is constructed in the countryside and the methods used by distant and local actors in generating and resisting change (Harvey, 1996). The past 15 years have been marked by the development of new 'market making' in rural areas, stimulated by deregulation and boom-and-bust economic cycles, not least in the housing sector. These processes have been a major agent of change for rural space, continually forcing productivist agriculture onto the back foot. The continuity and succession of family farming have come under particular threat, especially because farmers' sons and daughters are finding it more difficult to resist the attractiveness of seeing their family's land and buildings as multifaceted capital assets rather than as productive assets. With the emphasis on maintaining the mobility of capital and reducing sunk costs and the risks to investment, there is greater emphasis on flexibility of ownership and innovative forms of property relations (e.g. options, short-term leases and contracts, novel and divisible rights).

Many of the traditional rural interest groups have eventually conceded to these processes. For instance, the Country Landowners' Association, eager to protect its members' freehold property rights in the face of the demise of the agricultural priority and the growth in demand for rural amenity, now actively promotes the concept of rural enterprise zones, areas like their urban counterparts, which are free from planning control and provide entrepreneurs with developmental incentives. More generally, the recent batch of rural White Papers (and particularly the English version) constantly reminds the reader of the need to equate good countryside with a working countryside. It is important to recognize, however, that the types and the general architecture of the market-making ruralities which have developed over the contemporary period have the common feature of using *particular* commodity-based methods to deliver rural development. In addition, and by no means unrelated, these developments have tended to be privatized. That is, they have tended to ratchet up the social exclusivity of rural space and activities, whether concerning social housing, golf courses, or rural theme parks. This has been reinforced by policy makers and academics and their use of contingent valuation techniques, which give the impression at least of being able to suggest a price for any type of rural good or service.

Commoditization is useful at two levels of analysis. At the micro level it provides a way of examining actor strategies of resistance, accommodation and development; at the more macro level it has held a particularly dominant resonance in the redefinition of rural space as market-making space. At this level, many of the traditional rural organizations, such as the Countryside Commission and the Council for the Protection of Rural England, have been less sanguine about the progressive marketization of Rural UK plc. But very few alternatives have so far been developed to indicate how rural areas may cope with the new economic demands unevenly placed upon them (Tarling *et al.*, 1993).

Networks and actor spaces

The experience of the postwar agricultural productivism, and more recently of a prolonged period of uncertainty and restructuring in rural areas, has played havoc with established models of economic growth and development in rural space (Lowe *et al.*, 1995). Although the locality debates of the 1980s gave rural researchers a theoretical basis from which to focus more specifically upon the internal nature of local social economic processes, it has only recently progressed beyond generalized debates about rural development without giving precedence to either exogenous or endogenous derived development models (van der Ploeg and van Dijk, 1995). The construction and reconstruction of networks of social action and power transcend localities as well as being partly defined by them. A focus upon network construction and mediation thus begins to provide a way of breaking down the inevitable rigidities in conceiving rural space as derived from its physical composition or its strictly internal or external definition alone. This begins to collapse not only the exogenous and endogenous dichotomy but also the macro–micro problem. Murdoch and Marsden (1995) argue:

> Through the processes of association it is possible to do things in one place (the centre?) that dominate another place (the periphery?). The question of scale can therefore be posed in the following terms: what links local actors to non-local actors (i.e. actors in another locale) and how do these non-local actors effect change and control from a distance? It is through associations, and the ability they give certain actors to 'act at a distance', that actors-in-contexts, or as we prefer 'actor spaces', are tied together. (p. 372)

From this perspective it becomes necessary to examine the social and economic practices of actors in their context of action, or *actor space*. This involves assembling knowledge, resources, materials and technologies in order to progress actions and strategies, whether associated with a preservationist group or a corporate development firm strategically planning development proposals and options. The term *actor space* thus draws attention to the complexity of linkages between the material, phenomenological and social components of situations that are mobilized during the building of associations. It gives emphasis to the spatialization of action processes. If they are to progress to a position of changing economic conditions (perhaps through the development of new industrial units or the establishment of new recreational facilities in villages), these actor spaces need to become aligned with one another. This notion of alignment is derived from a concept proposed by Callon (1986) and from Latour's (1987) concept of 'actor network'. In this conceptualization, actors in discrete situations become bound into wider sets of relations which then alter the nature of individual actor spaces in accordance with the needs of the network as a whole (Law, 1992). The task, therefore, is to trace the formation, development and content of these networks, assessing how they can lead to developmental change, or seeing how constraints in actor spaces and in their alignments serve to restrict their development.

Given the earlier conceptual device of viewing rural restructuring from vertical (food) and lateral (non-agricultural) standpoints, it can now be seen how rural spaces are constituted and remade by cross-cutting networks of power and association, with rural restructuring as an outcome of the aggregated network effects. This is at the heart of rural *restructuring*. It is a purposive and forward-looking contingency. As Long and van der Ploeg (1995) argue:

> It is through the complex encounter and mediation of actors' projects that modes of ordering, that is specific routes to the future, are generated. The emergence of such ordering processes is the outcome of the interplay of different self-reflexive strategies, or what we have designated interlocking actors' projects. It is in this sense that we conceptualise structure. From the point of view of any one of the actors implicated in it, such a structure consists of a network of enabling and constraining entities (both human and 'delegated' non-human such as documents, machines, technology, and stocks of capital and material resources) and is therefore internally heterogeneous. That is, it is multiply composed and looks and functions for actors situated, as it were, at different locations and adopting different stances within the social landscape. (p. 69)

Such a theoretical position provides important opportunities for comparative empirical investigation. There are some interesting cross-national parallels. In Italy 'local rural systems' approaches, mainly concentrating on the formation of new food 'circuits', are demonstrating the significance of embedded social networks as a forerunner for local rural economic development (Picchi, 1994; Fanfani, 1994; Iacoponi *et al.*, 1995). Indeed, as the Common Agricultural Policy increasingly becomes cornered by trade liberalization tendencies on the one hand and pressures to deintensify agriculture on the other, more reliance will be placed upon assuming a positive relationship from the existence of dense social and economic networks and economic development gains in line with the somewhat mythologized Italian 'Third Italy' model. This is already being promoted by EU structures funding which expects 'partnerships' and locally developed networks to be in place as animators of local economic development.

The reality is obviously much more complex and variable, and one should be cautious in assuming that the 'density' or richness of extant networks necessarily leads to positive economic growth trajectories. Nevertheless, a network and actor-space approach, if developed though intensive research, does have the capability of progressing an understanding of rural social and economic change, as recent commentators have concluded (Murdoch and Morgan, 1996; van der Ploeg and Saccomandi, 1995). It also, interestingly, begins to expose rural areas as potential sites for economic innovation and new forms of exchange and transaction, associ-ated potentially with more sustainable food and environmental systems. This gives new forms of rural development an infusion of contingency and possibility which they have for so long seemed to lack. Moreover, from an analytical point of view, the focus on the content, power and contingency of network construction does not necessarily imply a populistic retreat to localism in the face of the ravages of harsh global forces. As many scholars, including social anthropologists, are aware, it is through a more sophisticated analysis of the 'local' that a broader comparative analysis of capitalism and globalization can be built. One does not exclude the other. For instance, in a telling passage, Miller (1995) has recently argued in a discussion on globalization that

> such ethnographic observations are vital if terms such as 'post-fordism' are ever to be more than glib generalities. Our micro-studies of consumption are not a retreat from political economy because we are finding that the local has become the command-ing heights of the political economy. It is in here that we can relate directly to ques-tions of, for example the comparative experience of World Bank sponsored structural adjustment. (p. 10)

And, as Feierman (1990) suggests:

> The wider world is not external to the local community, it is at the heart of the community's internal processes of differentiation. (p. 36)

DIFFERENTIATION AND COHERENCE IN RURAL CHANGE

Earlier approaches in the political economy tradition in the 1980s placed great emphasis upon understanding agriculture as a subset of the international food sys-tem, with rural development as an adjunct to these processes (Goodman and Watts,

1994; Marsden *et al.*, 1996). Since that period, scholars have begun to appreciate the growing centrality of rural space after the demise of the postwar compromise and have begun to tackle the diversity of rural space from a more nuanced social and political economy perspective. This chapter has examined some of the conceptual dimensions necessary to progress this development. They are dimensions which open the analysis of rural space to broader social science endeavours associated with the questions of the state, consumption and the possibilities for sustainable forms of production and consumption. Such progress needs to be founded on a process of carefully integrating aspects of nature and consumption into analyses. Perhaps easy to suggest, but it demands further conceptual refinement and specification.

Over the recent past many rural areas in Britain, as well as many parts of mainland Europe, have seen considerable growth in new employment and capital investment. Higher rates of new firm formation have occurred, particularly in the more accessible rural areas compared with urban areas. Such economic generalities serve, however, to mask the variable and socially active forces which contribute to these aggregated changes. The three key conceptual dynamics outlined in the second half of the chapter (regulation, commoditization and actor spaces) give some theoretical pointers to the ways in which more effective comparative analyses of rural economic change can be progressed, both within nation states and internationally. Such an endeavour becomes all the more important given the differentiation currently occurring in the development trajectories of rural space. Under the 'post-productivist' conditions now prevailing, and moreover, likely to continue, it is particularly relevant to consider how social, political and economic relations and outcomes become spatialized. Critical here is to appreciate how economic connections, distances and territorial arrangements are constructed, and how they are made and become active ingredients in the dynamics which help to create the new rural differences.

REFERENCES

Abram, S., Murdoch, J. and Marsden, T.K. (1996) *The social construction of middle England*. Papers in Environmental Planning Research 4, Department of City and Regional Planning, University of Wales, Cardiff.

Amin, A. and Thrift, N. (1995) Institutional issues for the European regions: from markets and plans to socio-economics and powers of association. *Economy and Society*, **24**, 41–66.

Chisholm, M. (1962) *Rural settlement and land use: an essay in location*. Hutchinson, London.

Callon, M. (1986) Some elements of a sociology of translation. In Law, J. (ed) *Power, action and belief: a new sociology of knowledge*? Sociological Review Monograph 32. Routledge & Kegan Paul, London.

Clark, G. (1992) 'Real regulation': the administrative state. *Environment and Planning A*, **24**, 615–27.

Cloke, P. and Thrift, N. (1990) Class and change in rural Britain. In Marsden, T.K., Lowe, P. and Whatmore, S. (eds) *Rural restructuring: global processes and local responses*, Critical Perspectives on Rural Change 1. Wiley, London, pp. 165–81.

Cochrane, W.W. (1968) *Farm prices: myth and reality*. University of Minnesota Press, Minneapolis MN.

Dexter, K. (1977) The impact of technology on the political economy of agriculture. *Journal of Agricultural Economics*, **28**, 211–21.

Fanfani, R. (1994) Agricultural change and agro-food districts in Italy. In Symes, D. and Jansen, A.J. (eds) *Agricultural restructuring and rural change in Europe*. Agricultural University Wageningen, Wageningen.

Feierman, S. (1990) *Peasant intellectuals*. University of Wisconsin Press, Madison WI.

Flynn, A. and Lowe, P. (1994) Local politics and rural restructuring: the case of the contested countryside. In Symes, D. and Jansen, A.J. (eds) *Agricultural restructuring and rural change in Europe*. Agricultural University Wageningen, Wageningen.

Flynn, A. and Marsden, T.K. (1992) Food regulation in a period of agricultural retreat: the British experience. *Geoforum*, **23**(1), 85–93.

Flynn, A. and Marsden, T.K. (1995) Rural change, regulation and sustainability. *Environment and Planning A*, **27**(8), 1180–93.

Goodman, D. and Watts, M. (1994) Reconfiguring the rural or fording the divide? Capitalist restructuring and the global agro-food system. *Journal of Peasant Studies*, **22**, 1–49.

Grabher, G. (ed) (1993) *The embedded firm: on the socio-economics of industrial networks*. Routledge, London.

Hall, P., Thomas, R., Gracey, H. and Drewelt, R. (1972) *The containment of urban England*. Allen and Unwin, London.

Harvey, D. (1996) The role of markets in the rural economy. In Allanson, P. and Whitby, M. (eds) *The rural economy and the British countryside*. Earthscan, London, pp. 19–40.

Hirsch, F. (1978) *The limits to growth*. Harvard University Press, Cambridge MA.

Iacoponi, L., Brunori, G. and Rovai, M. (1995) Endogenous development and the agro-industrial district. In van der Ploeg, J.D. and van Dijk, G. (eds) *Beyond modernization*. van Gorcum, Assen, The Netherlands.

Kneale, J., Lowe, P. and Marsden, T.K. (1991) *The conversion of agricultural buildings*. ESRC Countryside Change Initiative Working Paper 29, University of Newcastle.

Latour, B. (1987) *Science in action: how to follow scientists and engineers through society*. Open University Press, Milton Keynes.

Law, J. (1992) Notes on the theory of the actor-network: ordering, strategy and heterogeneity. *Systems Practice*, **5**(4), 379–93.

Long, N. and van der Ploeg, J.D. (1995) Reflections on agency, ordering the future, and planning. In Frerks, G. and den Ouden, J.H.B. (eds) *In search of the middle ground: essays on the sociology of planned development*. Agricultural University Wageningen, Wageningen.

Lowe, P., Murdoch, J., Marsden, T., Munton, R. and Flynn, A. (1993) Regulating the new rural spaces: issues arising from the uneven development of rural land. *Journal of Rural Studies*, **9**, 205–22.

Lowe, P., Murdoch, J. and Ward, N. (1995) Networks in rural development: beyond exogenous and endogenous models. In van der Ploeg, J.D. and van Dijk, G. (eds) *Beyond modernization*. van Gorcum, Assen, The Netherlands.

Marsden, T.K. (1995) Beyond agriculture? Regulating the new rural spaces. *Journal of Rural Studies*, **11**, 285–96.

Marsden, T.K. and Wrigley, N. (1995) Regulation, retailing and consumption. *Environment and Planning A*, **27**, 1899–1912.

Marsden, T.K. and Wrigley, N. (1996) Retailing, the food system and the regulatory state. In Wrigley, N. and Lowe, M. (eds) *Retailing, consumption and capital: towards the new retail geography*, Longman, London.

Marsden, T.K., Munton, R., Whatmore, S. and Little, J. (1986) Towards a political economy of capitalist agriculture: a British perspective. *International Journal of Urban and Rural Research*, **10**, 498–521.

Marsden, T.K., Lowe, P. and Whatmore, S. (eds) (1990) *Rural restructuring: global processes and local responses*, Critical Perspectives on Rural Change 1. Wiley, London.

Marsden, T.K., Murdoch, J., Lowe, P., Munton, R. and Flynn, A. (1993) *Constructing the countryside*. UCL Press, London.

Marsden, T.K., Munton, R., Ward, N. and Whatmore, S. (1996) Agricultural geography and the political economy approach: a review. *Economic Geography*, **72**, 361–75.

Miller, D. (1995) *Worlds apart: modernity through the prism of the local*. Routledge, London.

Mormont, M. (1990) Who is rural? Or, how to be rural. Towards a sociology of the rural. In Marsden, T.K., Lowe, P. and Whatmore, S. (eds) *Rural restructuring: global processes and local responses*, Critical Perspectives on Rural Change 1. Wiley, London.

Munton, R. (1995) Regulating rural change: property rights, economy and environment – a case study from Cumbria, UK. *Journal of Rural Studies*, **11**(3), 267–84.

Murdoch, J. and Marsden, T.K. (1994) *Reconstituting rurality: class, community and power in the development process*. UCL Press, London.

Murdoch, J. and Marsden, T.K. (1995) The spatialization of politics: local and national actor-spaces in environmental conflict. *Transactions of the Institute of British Geographers*, **20**(3), 368–80.

Murdoch, J. and Morgan, K. (1996) *Exploring the 'third way': networks in European rural development*. Report for the OECD, pp. 1–25.

Newby, H. (1979) *Green and pleasant land: social change in rural England*. Hutchinson, London.

Newby, H., Bell, C., Rose, D. and Saunders, P. (1978) *Property, paternalism and power: class and control in rural England*. Hutchinson, London.

Picchi, A. (1994) The relations between central and local powers as context for endogenous development. In van der Ploeg, J.D. and Long, A. (eds) *Born from within: practice and perspectives of endogenous development*. van Gorcum, Assen, The Netherlands.

Salais, R. and Storper, M. (1992) The four worlds of contemporary industry. *Cambridge Journal of Economics*, **16**(2), 169–93.

Tarling, R., Rhodes, J., North, J. and Broom, G. (1993) *The economy and rural England Strategy*. Review Topic Paper 4, Rural Development Commission, London.

van der Ploeg, J.D. and Saccomandi, V. (1995) On the impact of endogenous development in agriculture. In van der Ploeg, J.D. and van Dijk, G. (eds) *Beyond modernization*. van Gorcum, Assen, The Netherlands.

van der Ploeg, J.D. and van Dijk, G. (eds) (1995) *Beyond modernization*. van Gorcum, Assen, The Netherlands.

Ward, N. (1990) A preliminary analysis of the UK food chain. *Food Policy*, **15**, 439–41.

Ward, N., Lowe, P., Seymour, S. and Clark, J. (1995) Rural restructuring and the regulation of farm pollution. *Environment and Planning A*, **27**(8), 1193–1213.

Weber, M. (1949) *The methodology of the social sciences*. Free Press, New York.

Whatmore, S., Munton, R. and Marsden, T. (1990) The rural restructuring process: emerging diversions of property rights. *Regional Studies*, **24**, 235–45.

Winter, M. (1996) *Rural politics*. Routledge, London.

CHAPTER 3

SOCIAL PERSPECTIVES
Martin Phillips

INTRODUCTION: THE RECONFIGURATIONS OF RURAL SOCIAL GEOGRAPHY

The study of the social geography of the countryside has witnessed in recent years something of a reconfiguration in both its subject-matter and in the approaches its practitioners are adopting. One form of reconfiguration has been the re-examination of existing topics of research through a variety of new theoretical perspectives currently circulating, with seemingly increasing speed (Mohan, 1994), through the social and cultural sciences. So, for example, there have emerged new perspectives on such issues as rural demographic change and urban-to-rural migration, rural resources and service provision in the countryside, and the nature of rural communities.

Another reconfiguration has been the emergence of new themes in the social geography of the countryside, relating to issues of social identity, social difference and the construction and reception of cultural images of the countryside. This reconfiguration has likewise involved the bringing of social and cultural theories into rural social geography, although to some limited degree it may also have positioned rural geography more centrally within social and cultural studies. The aim of this chapter is to outline these two strands of the reconfiguration of social geographical perspectives on the countryside, to illustrate some examples of work within what might be termed a 'new rural social geography', and to explain some of the issues that the reconfiguration of rural social geography raises for students of the rural. The chapter will begin with a brief survey of the character of rural social geography before these changes took place.

RURAL SOCIAL GEOGRAPHIES UP TO THE 1980s

For many people the term *social geography* has referred principally to the enumeration of people living in rural areas. In the 1950s and 1960s, for example, the focus of much of rural social geography was on the issue of rural depopulation, whereas from the 1960s this was seemingly replaced by a concern with counterurbanization, or the movement of people from towns to the countryside. Considerable attention has been paid by geographers and others to documenting the extent of depopulation and counterurbanization (e.g. Berry, 1976; Bolton and Chalkley, 1990; Brown and Wardwell, 1980; Champion, 1989; Fielding, 1982; Frey, 1987; Halliday and Coombes, 1995; Lawton, 1968; McCarthy and Morrison, 1977; Woodruffe, 1976). This empirical focus became closely associated in the 1970s with a logical positivist search for some more or less universal law with which to explain the observed changes.

So, for example, depopulation was seen as the result of economies of scale and cumulative advantage of the process of centralization, whereas counterurbanization was explained as the outcome of such laws as the maximization of individual preferences, economic cost minimization and the balancing of demand and supply (Fielding, 1982). Rarely were such accounts explicit in their use of a logical positivist approach. Indeed, many studies remained either resolutely empirical, concentrating on enumerating the extent of population change rather than attempting to explain it, or focused on spatially and temporally contiguous observable events such as local success when implementing a development policy (e.g. Hill and Young, 1991; Parker, 1984; Pettigrew, 1987), local success in market competition (e.g. Drudy and Drudy, 1979; McCleary, 1991; Strachan, 1988), and the existence of people with the resources and desire to live in the countryside (Bolton and Chalkley, 1990; Joseph *et al.*, 1988; Joseph *et al.*, 1989).

Two other important social foci for rural geographers in the 1970s and early 1980s were, first, resource conflicts and their 'management' (e.g. Cloke and Park, 1985; Coppock and Duffield, 1975; Patmore, 1983); and second, people's access or lack of access to services (e.g. Clark, 1982; Moseley, 1979; Phillips and Williams, 1982; Shucksmith, 1981). These studies drew to varying extents, and generally implicitly, on the philosophies of empiricism and logical positivism. In addition many of them drew, often rather more explicitly, on the notion of applied geography; that is, geography oriented to solving social problems. One of the clearest calls for such an approach was made by Gilg, who concluded his *Introduction to Rural Geography* with the statement:

> The future for rural geography should be an applied one, where it integrates its own research, relates this to the real behavioural world and to policy formation, and thus attempts to produce a rural environment that is not only physically attractive but also a lively and prosperous place to live. (Gilg, 1985, p. 266)

Cloke (1989a, p. 167) gives a useful list of works which endeavoured to provide both academics and planners with information about the contemporary character of rural areas.

A fourth focus of rural social analysis was on the 'rural community' and the changes affecting these communities. Harper (1989) has commented that many of the studies of rural communities by geographers in the 1950s and 1960s were characterized by an 'abstracted empiricism' in which there was no attempt to outline explanation independently of the description of events and no element of comparison between case studies. Many studies drew, albeit often implicitly, on the theoretical arguments of the German sociologist Ferdinand Tönnies and his distinction between *Gemeinschaft* and *Gesellschaft* (Tönnies, 1957). Tönnies argued that there were two basic types of human relations: (1) *Gemeinschaft* 'community' relations based on 'close human relationships developed through kinship . . . common habitat and . . . co-operation and co-ordinated action for social good,' and (2) *Gesellschaft* 'society' relations created through 'impersonal ties and relationships based on formal exchange and contract' in which 'no actions . . . manifest the will or spirit of . . . unity' (Harper, 1989, pp. 162–63). According to Tönnies, and more especially later writers who have drawn upon his ideas, these social relations were linked to a spatial division between urban and rural space. Rural areas were frequently described as places of community or *Gemeinschaft* and urban places were linked places of impersonal society or *Gesellschaft*.

Although such arguments were often utilized implicitly within empirical case studies, there emerged particularly in the 1960s and 1970s three areas of more explicit theoretical debate. The first was over the degree to which there was a rural–urban dichotomy in the two social relations or whether there was a rural–urban continuum (cf. Frankenberg, 1966; Redfield, 1947; Wirth, 1938). The increasing acceptance of a rural–urban continuum was radically disrupted by the work of Pahl (1966), who argued the untenability of the notion that particular forms of social relationships were related to spatial units such as town and countryside. He suggested that 'geographically tainted concepts' such as urban societies and rural communities should be replaced with more 'sociological concepts' of 'national' and 'local' ways of life, which he saw as relating to degrees of freedom from social constraints and thereby to sociological factors like social class and stage in the life cycle.

The research agenda for Pahl was therefore to examine the degree to which particular places were inhabited by people with local and national lifestyles, an argument which was well demonstrated in his own study of villages in Hertfordshire (Pahl, 1965). The third area of debate focused on temporal change and whether there was a transformation over time in social relationships within particular settlements. Although Pahl preferred to talk about these social relations using his sociological concepts of local and national ways of life, other studies used either the distinctions of Tönnies or such terms as rural and urban ways of life. What remained central to many of these studies was what Williams (1985, p. 96) has described as one of the most powerful of 'modern myths', in which social changes such as industrialization and urbanization are seen to bring about 'a fall' or decline in the character of society.

This review of the four foci of rural social geography – as the study of population change, access to resources, access to social services, and changes in community life – has necessarily been very brief. But do note the existence of important differences within and between these four foci; those wishing to explore these aspects in more detail might usefully look at texts such as Lewis (1979), Harper (1989) and Robinson (1990). It can, however, be argued (Phillips, 1997) that they share what Philo (1992a, p. 3) has called a 'restricted social imagination' in that they tend to avoid considering phenomena which are *immaterial*, that is which are not tangible, not easily countable, and which have 'worrying *political overtones*'. During the course of the 1980s and 1990s these restrictions on the social imagination of rural geographers have come under challenge; in part, this relates to rural studies becoming connected into ideas circulating across the wider social sciences. In particular, many rural geographers have come to recognize the political overtones of studying rural social geography through incorporating some ideas from Marxist and neo-Marxist political economy, and to see the immateriality of rurality as being of considerable importance through bringing in ideas from post-modernism, post-structuralism and cultural studies. As mentioned earlier, this incorporation of ideas circulating within the wider social science community has affected the study of existing 'objects of research' within rural social geography and led to the study of a number of new social phenomena. The rest of the chapter will seek to outline both the incorporation of 'political-economy' and 'post-modernism/post-structuralism/ cultural studies' perspectives into rural social geography and their impacts on existing and new objects of research.

MARXISM AND THE POLITICIZATION OF RURAL SOCIAL GEOGRAPHY

The idea of critical rural studies

The issue of the political dimensions to rural studies became particularly significant in the late 1970s and 1980s when a number of rural researchers began to argue for the adoption of a 'radical' or 'critical' approach (e.g. Newby, 1977; Newby and Butler, 1980; Hoggart, 1987; see also Cloke, 1989a; Phillips, 1994). The adjective *critical* was used to imply a 'more independent and sceptical attitude towards rural phenomena' (Newby and Buttel, 1980, p. 2), which 'was prepared to reject the accepted structures, institutions and perceptions of the countryside' (Phillips, 1994, p. 89).

The significance of this claim can be demonstrated by returning to Gilg's views about an applied rural geography being concerned with producing a countryside that is physically attractive, lively and prosperous. Although this may seem a laudable aim, as Hoggart and Buller (1987, p. 266) pointed out, it assumes there is some agreement over what constitutes a physically attractive, lively and prosperous environment. The dangers of making such an assumption have been clearly demonstrated in Harrison's (1991) study of recreational visits to the countryside. She suggests that much of rural recreation planning is based on a *countryside aesthetic* which 'portrays the countryside as fine landscape and its appropriate enjoyment as being achieved through solitary and quiet pursuits' (Harrison, 1991, p. 2). This very much echoes the sentiments of Gilg about the need for a physically attractive landscape, but Harrison found that many people did not go into the countryside to merely enjoy an attractive physical landscape but rather went to experience 'a range of active and sensual pleasures'.

Instead of a singular view of the countryside, people appear to have multiple and indeed often seemingly contradictory feelings and perceptions about it. Deciding upon what to change in the rural environment is much harder if you start recognizing a variety of different, and perhaps incompatible, viewpoints about what the countryside is and should be. In practice, Harrison argues, these different and complex views are ignored because they do not fit in with the perceptions of planners and landowners who structure the provision of countryside recreation. In other words, the views of applied rural geographers tend to mirror the views of those who have the most power to construct the countryside rather than reflect a social consensus on the use of the countryside.

Taking on board such arguments raises a number of important questions about the study of rural geography. In particular, recognition of other views and the extent of their marginalization politicizes the study of the countryside: actions to make the landscape more physically appealing may be in the interest of one group but not another, and furthermore, they may actually favour those who are already getting more of their interests served anyway. This is why rural researchers such as Newby and Buttel argued it was necessary to examine and sometimes to reject the generally accepted views of what the countryside is and should be, and in accounts of the rural, to override the subjective experience of individual people (Newby and Buttel, 1980). Linking to more general accounts of philosophy in human geography, the argument was to avoid overly 'voluntarist accounts' of human actions and to recognize the influence of social structures which lay beyond the individual and the discursive (e.g. Gregory, 1978, 1981). In both human and rural geography, the advocacy of a critical approach was generally quickly followed by the adoption of a variant of Marxist or neo-Marxist political economy (Cloke, 1989a; Cloke and Little, 1990).

The notion of critical rural studies and the adoption of a political economic approach became particularly significant within the study of agricultural change (Chapters 2 and 4). They have also affected the study of rural planning and policy making (Cloke, 1987; 1989b; Cloke and Little, 1990) and, albeit quite slowly (Cloke and Little, 1990), the four areas of rural social geography identified above.

Political economic approaches to rural social geography

One of the most significant features of the adoption of a Marxist political economy for rural social geography was that it suggested links between the four areas of rural social geography. The study of rural population change, resource conflict and management, access to services, and rural community change all effectively became the study of the outcome of relations, structures and agents of political economy. For the sake of brevity, attention will focus in this section on the impact of political economy perspectives on the study of rural demographic change and rural communities, but broadly similar arguments could be advanced for the study of rural resources and services (Chapter 11).

With regard to rural demographic and community change, one of the key conceptual arguments was to suggest these changes had an important, and hitherto rather neglected, class dimension. Cloke and Thrift (1990, p. 165) suggested that much of rural studies exhibited a rural ideology which had an 'aversion to notions of class' and which portrayed rural society as 'essentially classless . . . even if . . . unequal and hierarchical'. Proponents of this rural ideology were able to draw theoretical legitimation from both the *Gemeinschaft–Gesellschaft* distinction, and the 'invisibility' of rural issues in political economy and class theory (Hamilton, 1990; Murdoch and Marsden, 1994, p. 1; Winter, 1984). However, as Murdoch and Marsden (1994) have remarked, several rural researchers have examined class in the countryside.

It is indeed possible to identify within rural studies at least five important approaches to class analysis, all of which draw to some extent on Marxist political economy. First, researchers such as Newby *et al.* (1978) and Buttel and Newby (1980) argued that rural class relations were formed around property relations. The general thrust of their argument has been clearly spelt out by Newby, who suggested that it was possible

> to follow a chain of events which led from the continual reorganisation of property relations in agriculture . . . through to changes in social composition of rural areas and on into an analysis of emergent social conflicts in the countryside, of which issues relating to environmental conservation, employment growth, and housing may be regarded as emblematic. (Newby, 1986, p. 212)

A clear illustration of this argument is provided by Newby (1987), who developed a social history of rural England centred around four property systems. The emphasis on property is also evident in the recent studies associated with the ESRC Countryside Change Initiative (e.g. Marsden *et al.*, 1993; Munton, 1995; Murdoch and Marsden, 1994; Whatmore *et al.*, 1992).

Authors such as Bradley (1981), Barlow (1986), Rees (1984) and Urry (1984) criticized this approach, arguing that property relations were as much socially created as the social phenomena they were supposed to explain. Barlow (1986, p. 311), for example, argued that property relations were 'both a cause and an outcome of class and social struggles', whereas Rees (1984) suggested that while agrarian property relations may have been of considerable significance in the past, the present rural population is affected by a wider range of processes. The idea emerged that rural social change was essentially 'capital driven'; that is, it was the outcome of changes created by the need of capitalist industries to maintain profitability. Particular stress was placed on the restructuring of employment, and Rees has put forward the following argument:

> Changes in rural employment structures are central to any understanding of the reality of rural social life. On the one hand they reflect profound shifts in the nature and organisation of capitalist production and, more specifically, the widely differing types of locality. On the other, employment changes themselves have resulted in radical developments in terms of rural class structures, gender divisions, the forms of political conflict occurring in rural areas and, indeed, of the complex processes by which 'rural cultures' are produced and reproduced. (Rees, 1984, 27)

The shift from rural depopulation to counterurbanization was interpreted as the result of an urban-to-rural shift in industry, itself seen as the result of reorganization of a capitalist economy. The nature of this approach is clearly illustrated by Rees (1984) and by Day *et al.* (1989), who argued that the character of rural mid-Wales, including its population densities, could be understood as the outcome of shifts in the organization of capitalist production (Table 3.1).

Table 3.1 Capital-led social change in rural mid-Wales

Shifts capitalist mode of production	Spatial focus of capital investment	Impact on rural mid-Wales	Period of existence
Rise of industrial capitalism	Formerly rural areas in north of England, east Midlands, south Wales and Scotland These areas grew into urban areas	Decline of rural industry in mid-Wales and increasing reliance on agrarian petty capitalism Draining of people and capital from rural Wales to fuel new industrial developments, particularly in south Wales; depopulation	18th to 19th centuries
Rise of Fordist production	Industrial regions in London and west Midlands, prosperous agricultural regions such as East Anglia	Agricultural decline and depopulation	1900s to 1960s
New industrial division of labour	Establishment of branch plants in local areas with cheap labour and headquarters in areas of skilled labour	Establishment of externally controlled industries, i.e. branch plants, sustaining rural population	1960s to 1970s
Post-Fordist production or 'disorganized' capitalism	Variety of alternative strategies	Unclear (but see Cloke and Goodwin, 1992)	1980s

Source: Adapted from Day *et al.* (1989).

For several authors, the shift from property to occupation was still insufficient to capture the dynamics of contemporary class relations. In particular, a 'restructuring approach' has emerged that seeks to integrate the relations of property and occupation with those of consumption and commodification (Cloke, 1993a; Lowe *et al.*, 1993; Marsden *et al.*, 1993). As Lovering (1989, p. 198) records, the term *restructuring* is used in a wide range of contexts but generally implies 'qualitative changes in the relations between constituent parts'. Within rural political economy studies, one of the key changes which has been identified is the relationship between production and consumption. In particular, it has been argued that rural areas have increasingly become spaces of consumption as opposed to spaces of production (Chapter 2). One aspect claimed for this restructuring is that formation of social class is 'disarticulated from being "grounded" in production' (Murdoch and Marsden, 1994, p. 7) and there is a transfer of 'the perceived fulcrum of rural tension from the realm of production to that of consumption' (Miller, 1996, p. 8). Attention is directed at the way particular social groups desire particular forms of rurality and how they act, both individually and collectively, to achieve their own rural idylls (Marsden and Murdoch, 1990; Murdoch and Marsden, 1994; 1995).

A fourth political economic perspective applied to the study of rural social phenomena such as population change and community relations is regulationist theory (Cloke and Goodwin, 1992; Goodwin *et al.*, 1995; Marsden, 1995). Although there has been considerable debate about the precise characteristics and status of this theory (e.g. Jessop, 1990; Murdoch, 1995a; Tickell and Peck, 1992), one of its key starting-points is an attempt to develop a political economic perspective which avoids the functionalism and economicism seen to be associated with structural Marxism. In particular, it has been argued that regulationist theory seeks to understand the links between 'the political economy of rural change and the concurrent cultural representations, political strategies, and social conflicts' (Goodwin *et al.*, 1995, p. 1245). The approach has close connections with restructuring theory, and indeed the terms are often used interchangeably. Regulationist theory tends to emphasize how people and social agencies deal with and seek to direct economic change, whereas many of the restructuring theorists adopt a more structural–functional form of analysis. Regulationist theory has not been without its critics; one rural researcher argues:

> Despite efforts to move regulation theory away from the totalizing discourse of structural Marxism, in the end the struggle to introduce contingency, indeterminacy, and process merely serves to highlight how far the ensuing account is tied to its initial assumptions. (Murdoch, 1995a, p. 737)

Closely connected with restructuring and regulationist perspectives on rural change has been class analysis, in which class relations and class groupings are seen as concurrently the end product of past rounds of economic restructuring and an agent of the remoulding of ensuing cycles of restructuring. A particularly clear illustration of this argument is the work of Cloke and Thrift (Cloke and Thrift, 1987; 1990; Thrift, 1987), which shares with the economic restructuring and regulation approaches a recognition that rural changes are connected to more general economic transformations. In particular, it is suggested that contemporary rural change is linked to a shift from a manufacturing-centred economy to a more service-centred economy, in which capitalist–working class relations based on ownership of capital and labour are overlain by social relations based on such things as skills and qualifications, consumption decisions and political power created through corporations and state bureaucracies. These new social relations are seen to lead to new sources of social power in addition to those produced from the capital–labour relation, and to the emergence of a new service class able to utilize these new sources of power. This new class has both the power and the desire to live in the countryside (it can therefore be seen as a major coloniser of the countryside) and once living there it changes, restructures or reconstitutes the countryside.

Sometimes this restructuring is quite direct: Thrift (1987) writes of the service class 'covering' their homes with Laura Ashley prints and stripped pine furniture, while Murdoch and Marsden (1991) and Cloke *et al.* (1991) have described attempts to 'manufacture' desirable rural homes which might appeal to executives and other members of the service class. Other impacts are more circuitous, involving a series of iterations between the service class and the social and economic structure of places. Cloke and Thrift (1987) argue that early service-class migrants can influence later class colonization, either positively by expanding work, housing, communication and consumption in an area, or negatively by seeking to conserve the area from the impact of more colonization. This feature is related to the notion of 'nimbyism', whereby recent rural in-migrants seek to prevent further developments in their backyards, even though they themselves may be living in a recent development.

An important theme of this class analysis has been the issue of conflict; Cloke and Thrift (1987; 1990) as well as Cloke and Little (1990) have suggested that much rural social research has misinterpreted the basis of rural social conflict. In particular, they criticize the notion that rural social conflict necessarily stems from middle-class incomers disrupting an established geographical community by displacing working-class locals. At least three problems with such an interpretation can be identified. First, many of those considered to be 'local' will be in the middle or higher social classes; classic examples are farmers, which Newby *et al.* (1978) see as the archetypal petite bourgeoisie. Second, in-migration in many areas has been going on for such a long time that one may well be witnessing middle-class replacement of other middle-class residents, rather than the middle-class replacement of a working-class population. There is indeed a growing theoretical literature which suggests that the middle class is far from uniform (e.g. Bourdieu, 1984; Goldthorpe, 1982; Gouldner, 1979; Savage *et al.*, 1988, 1992; Wright, 1978, 1979, 1985) and Cloke and Thrift (1990) have suggested that there may be a series of 'tournaments of taste' and even out-right conflict between middle class fractions. Third, there is some evidence that not all contemporary rural colonists are middle class: Cloke *et al.* (1994) have reported that several rural areas in England had 'recently received significant proportions of lower income in-migrants,' while Cloke, Phillips and Thrift (Cloke *et al.*, 1991; Cloke *et al.*, 1995, 1997; Phillips, 1993) have identified a range of distinct 'channels of entry' by which social classes without access to a well-paid occupation and/or long career paths can come to 'colonize' the countryside (Table 3.2).

The class analysis approach has come in for some criticism (Hoggart *et al.*, 1995; Miller, 1996; Murdoch and Marsden, 1994). One issue is whether the service class constitutes a coherent class or merely some class fraction. Hoggart *et al.* (1995) suggest that, despite 'assertions of service class sociopolitical and cultural dominance', there are also clear expressions of 'uncertainty over what the service class is'. More specifically, they suggest Cloke and Thrift's (1990) claim that the service class is itself fractured by non-class relations, such as differences of consumption, gender, public and private sector activity, life cycle and place, provides 'such a heady list' that it leads them to 'the conclusion that it is inappropriate to talk of *a* service class at all' (Hoggart *et al.*, 1995, p. 213, italics added). Another

Table 3.2 Contemporary channels of entry into rural areas and assets of class formation

Process of colonization channel of entry	Process of class formation	Illustrative class fraction
Mainstream	Good labour market position Stable career structure	Bourgeoisie Private sector professional Manager and government bureaucrat
Petit property dealing	Petit capital accumulation	Petite bourgeoisie (mercantile)
Marginal gentrification	Sweat equity	Petite bourgeoisie (craft)
Marginal dweller	Accommodation	Working and unemployed

Source: Adapted from Cloke *et al.* (1995).

argument subject to some critical comment is the claim that the service class has a particularly strong attachment to the countryside (Thrift, 1989). In particular, Savage *et al.* (1992, p. 104) suggest the claim is based on loosely collected impressions rather than on 'conceptually informed survey work'. They go on to examine the differences in the way social groups consume various products and activities on the basis of market research carried out by the British Market Research Bureau:

> Insofar as Thrift's emphasis upon the 'countryside orientation' of the 'service class' is borne out by our findings, it is largely the managers who appear to indulge in it. (Savage *et al.*, 1992, p. 116)

They elaborate the findings of their study into a suggestion that each of the middle classes they have identified has its own particular set of consumption practices and cultural principles (Table 3.3). Their arguments show close parallels with the work of Wynne (1990) and Harrison (1991), and they have been discussed in a rural context by Murdoch (1995b), Murdoch and Marsden (1994) and Urry (1995b). Urry, while broadly accepting the arguments of Savage *et al.*, suggests that consumption practices and cultural principles are not clearly class bounded. He concludes by suggesting that social changes taking place in rural areas, particularly those related to leisure and consumption, are 'certainly *related* to changes in wider society, including the striking growth of a service class, but that cannot be literally described as the service class remaking of rural localities' (Urry, 1995b, p. 215). This point is reinforced by recent arguments that the countryside may not only frequently be a middle-class territory, but also in many respects a racialized, nationalized, aged, sexualized and gendered space (Agg and Phillips, 1997; Cloke *et al.*, 1995; Murdoch, 1995b; Phillips, 1993).

Overall, one can suggest that rural class analysis has reached something of a dilemma, arguably even a hiatus. There has been a steady movement away from seeing class as being the outcome of a narrow set of relations centred in the workplace, towards a more all-encompassing view which, as Murdoch and Marsden

Table 3.3 Social class, consumption practices and cultural principles

Social group	Cultural principle or habitus	Consumption practices
Public sector welfare professional	Ascetic	Healthy and sporty living High culture
Private sector professionals and service workers	Post-modern	Healthy and extravagant living Mix of high and low culture No overarching organizing principle
Managers and government bureaucrats	Inconspicuous consumption	Low participation in high culture Preference for a cleaned-up countryside

Source: Adapted from Savage *et al.* (1992).

(1994, p. 17) have recently put it, sees class formation as involving activities that 'ostensibly have no class complexion'. It has even been argued that class is little more than an umbrella term for a set of 'disparate social processes and phenomena' (Murdoch, 1995b, p. 1214). On the other hand, Miller appears to remain wedded to a more classical Marxist political economy:

> Rural studies urgently need a clearer exposition of what this drift away from social relations of production implies for the latent but over-arching assumption of dialectical change and the dynamics of historically significant developments. (Miller, 1996, p. 10)

Miller's argument overplays the extent to which rural class analysis has drawn, even implicitly, upon a dialectical materialist form of political economy. It also ignores the series of criticisms which have been raised against an overly structuralist interpretation of class, both within class analysis (Goldthorpe and Marshall, 1992; Murdoch, 1995b; Pahl, 1989; Savage, 1994; Savage and Butler, 1995) and more generally related to the rise of post-modern and post-structuralist philosophies. It is to the challenge posed by these philosophies that attention now turns.

POST-MODERNISM AND POST-STRUCTURALISM IN RURAL GEOGRAPHY

The rise of post-modernism and post-structuralism in rural geography

The philosophies of post-modernism and post-structuralism have been circulating in social and particularly cultural studies for many years (Connor, 1989). Their explicit entrance into rural studies, however, had to wait until the early 1990s, when there was a series of exchanges about the nature of post-modernism and its relevance to rural studies (Cloke, 1993b, 1994; Halfacree, 1993; Murdoch and Pratt,

1993, 1994; Philo, 1992b, 1993). Particularly significant was an article by Chris Philo in the *Journal of Rural Studies*; entitled 'Neglected rural others', it argued, among other things, that much of rural geography had effectively been 'peopleless' in that it was

> written in such a way that rural landscapes are either deserted of people . . . or occupied by little armies of faceless, classless, sexless beings dutifully laying out Christaller's central place networks, doing exactly the right number of hours farmwork in each of Von Thünen's concentric rings, and basically obeying the great economic laws of minimising effort and cost in negotiating physical space. (Philo, 1992b, p. 201)

Here is a clear criticism of empiricist-influenced and particularly logical positivist-influenced rural geography. However, Philo's criticism of peoplelessness was not simply aimed towards these forms of rural geography but was also directed at many 'contributions to rural geography that stand outside of the spatial-scientific mode of treatment', including those which proceed from Marxist and political economy perspectives (Philo, 1992b, p. 200). He effectively claimed, as did Murdoch and Pratt (1993) in a further elaboration of the implications of a post-modernist approach to rural studies, that much of rural studies is 'modernist' in character in that it seeks to specify, using a few key theoretical claims, how the world, in general, operates. Philo argues that this modernist approach leads to the development of accounts of social life which 'inevitably steamroller over the more specific "stories" that "other" people in "other" places tell themselves when seeking to make sense of their specific and situated existence' (Philo, 1993, p. 198).

In the context of rural studies, Philo argues that one of the principal costs of the desire to construct all-encompassing theoretical accounts of rural social life and rural change has been the neglect of 'those people who stand outside of the societal "mainstream"' . . . who are not male, white, heterosexual, middle class, middle-aged, able-bodied and sound-minded' (Philo, 1993, p. 430). Philo suggests this is partly because the people who construct modernist theorizations are 'white, middle-class, middle-aged, able-bodied, sound-minded, heterosexual men,' often living in major urban centres, and partly because the theorizers tend to assume the people they study are like them, even the same. As a result, they tend to portray rural people as being like the rural researcher, 'men in employment, earning enough to live, white and probably English, straight and somehow without sexuality, able in body and sound in mind, and devoid of any other quirks of (say) religious belief or political affiliation' (Philo, 1992b, p. 200). The result is, Philo suggests, that much of rural geography appears to be populated by indistinct 'Mr Averages'.

Clearly, neither all rural researchers nor all the rural researched fulfil all the criteria of 'the Same'. Instead, there are myriad lines of difference within both populations. Post-modernists such as Philo argue there is a need to recognize, even celebrate, these differences. There have certainly emerged several subjects of rural social research related to the study of difference and neglected others. Bell and Valentine (1995) have discussed lesbian and gay lives in the countryside as portrayed in novels, poems, films and television programmes, and as acted out in the lives of rural dwellers and in rural recreation. In an important refraction of rural social geography as the study of access to resources, they also record how the lives of gays and lesbians in the countryside are affected by the lack of 'gay facilities and services' (D'Augelli and Hart, 1987; Moses and Buckner, 1980; Rounds, 1988).

Other new subjects of attention by rural social researchers include people of colour in the countryside and the traveller. With reference to the first group, Kinsman (1995) illustrates how many images of the British countryside are racialized in that they 'naturalize' the presence of white settlers and make representations of other groups appear unusual and in one way or another problematic. Similar arguments could be made about the traveller. Halfacree (1996) has highlighted how nomadism may contravene the dominant social representation of rural space as a 'rural idyll'. Particularly transgressive of the idyll, he suggests, are the New Age travellers. The term *New Age traveller* emerged in the late 1970s and grew in particular prominence from the mid-1980s. It is a label that has been applied to quite a diverse range of people including, at times, 'Gypsies and Irish and Scottish Travellers . . . ravers, hunt saboteurs and environmental protesters' (Sibley, 1995, pp. 106–7). The people labelled as New Age travellers are frequently portrayed as being very different from the more traditional rural settlers and visitors of the mainstream middle classes; they are thereby subject to a series of negative comments by politicians and within the mass media (Cloke, 1993b). In practice, however, these groups appear to share the same ideals about the countryside. Studies by Halfacree (1994a, 1994b), Hetherington (1995) and Urry (1995b) have suggested that both New Age travellers and the rural middle class show elements of what might be called *Gemeinschaft* in that they value emotional attachments to place and to an imagined and socialized community; both groups also exhibit tribal characteristics such as an emphasis on the symbolic over the material, the ritualization of living and a clear differentiation in behaviour between members and outsiders (Shields, 1992).

The issue of ambiguous identities and values has become of heightened significance in rural geography in relation to the rise of post-structuralist ideas. As Pratt (1994) notes, post-structuralism both draws on and extends structuralist ideas as they relate to language and the subject or person. Post-structuralists adopt what Habermas (1987) describes as a 'world generative' view of language, which sees it as having its own developmental logics having little or no connection to any extralinguistic or 'objective' reality. Terms and ideas are defined in relation to other terms or ideas rather than reflecting some relationship with an object. Halfacree (1993) has adopted such arguments in a discussion of definitions of rurality. He suggests that conceptualizations of rurality are not created as direct, mimetic, re-presentations of the character of particular places or localities but instead they are 'signs without a referent'. In particular, Halfacree argues that it is useful to consider rurality as a socio-spatialization, i.e. disembodied mental structures which focus on space and society and in which 'spatial metaphors and place images . . . convey a complex set of [social] associations' (Shields, 1991, p. 46). These cognitive structures underlie a variety of discourses or circulations of meaning, including those of academics and those of the people they study. This point has been elaborated by Jones (1995), who suggests that rural studies need to address not only academic discourses on rurality, but also popular (or media), professional, external (or inter-subjective) and internal (private or subjective) discourses.

One consequence of such arguments has been to lead rural social research towards an engagement with the cultural analysis of landscapes (Cosgrove and Daniels, 1988; Daniels, 1993) and discourses of 'nature' (Bell, 1994; Lawrence, 1995; Mormont, 1987; Wilson, 1992), of 'town and country' (Short, 1991; Williams, 1985), of 'village England' (Matless, 1994) and 'communities' (Short, 1991; Williams, 1985). Another consequence is an increasing use of post-structuralist and post-modernist sociocultural theorizations and concepts. Lawrence (1995, p. 302) and Cloke (1992; 1994), for example, have suggested that rural geographers might draw on the ideas of Baudrillard (1983a; 1983b) and examine how notions of rurality which appear to bear little relation to the actual conditions of many rural spaces actively influence how people act. Murdoch and Pratt (1993) have even proposed that the notion of rurality should be replaced by the idea of post-rurality, which does not relate to any standard definition of what the rural is, but does encompass any phenomena that any people take to be rural. It may also be relevant to talk of the countryside as a hyperreality in which the representations of rurality, even those accepted as being in important senses fictions, actively come to structure rural spaces. The use of Lyme Hall in Cheshire to portray Pemberley in the television adaptation of Jane Austen's *Pride and Prejudice* apparently led to a surge of visitors from some 800 per week to over 5500 in just two days (Ward, 1995).

The second important focus of concern within post-structuralism is the issue of 'the subject'. As Pratt puts it:

> Poststructuralists also absorbed the antihumanist critique of a unified, knowing and rational subject, instead interpreting subjectivity as continually in process, as a site of disunity, conflict and contradiction. (Pratt, 1994, p. 468)

The impacts of such ideas in rural geography have been twofold. First, with reference to methodology and epistemology, there has been an increasing rejection of the notion that 'true knowledge' can be obtained by adopting the perspective of a detached observer, unswayed by personal values or interests (Phillips, 1994). Instead there has been a recognition of the 'situatedness' and 'positionality' of knowledge claims. Researchers cannot find a privileged vantage point which is outside the society in which they live, and their view of a particular situation is always influenced by the position and person doing the observing, not least because 'researchers unconsciously determine the form of responses from those they encounter in the research process' (Phillips, 1994, p. 113). The implication of such arguments is that rural researchers should recognize the impossibility of being a completely detached objective observer; instead they should seek to reflect on how they connect with those they are researching and how those connections influence what knowledge the researcher gains.

A second implication of post-structuralist notions of the subject has been an increasing awareness of the fluidity and multidimensionality of people's social identity. Harrison (1991), Clarke *et al.* (1994) and Urry (1990; 1995b; 1995c) have argued that people may temporarily adopt new identities when they participate in leisure pursuits and in political protests. The quiet librarian during the week may engage in war games at the weekend, while the staunch Conservative voter and believer in the rule of law might take to chaining themselves to trees about to be felled to make way for a road bypass. Furthermore, these people might well be joined in their leisure pursuits and political actions by a range of quite different people. According to Urry (1995b, 1995c), the result is a breakdown of established lines of social differentiation and social interaction; the breakdown of established social differentiation is termed *detraditionalization* and the breakdown of established social interaction leads to *new socializations*, which are like communities in that people share a common identity and goals but are different in that people become members from choice (rather than as a consequence of say birth, occupation or place of residence).

Such arguments raise significant questions for, among other things, rural class analysis because they suggest that not only is a person's self-consciousness of their identity not tied necessarily to lines of social differentiation such as class, but also that people may act in non-class-centred social groups. For some commentators, such processes spell the demise of class:

> Class has all but disappeared from the mainstream of intellectual attention. All the attention now focuses on social divides such as gender and sexuality, ethnicity and ecological consumption, of which the 'cultural turn' in so many subjects in the social sciences and humanities is both a cause and a symptom. (Cloke *et al.*, 1995, p. 220)

Cloke *et al.* argue that the writing off of class is premature but that it is important to recognize how the significance of class cannot be automatically assumed (cf. Miller, 1996), how other social relations such as gender, sexuality, race and ethnicity may be of equal or greater significance in certain instances, and how it may even be that 'the determinants of class may have changed to such a degree that we no longer recognise them' (Cloke *et al.*, 1995, p. 222). Phillips (forthcoming) has argued for the exploration of what Savage (1994) and Savage and Butler (1995) have described as an 'interpretative approach to class' which recognizes 'the ambiguities of class, the difficulty of locating it in only one dimension, and the "messiness" of its connections with other social phenomena' (Savage and Butler, 1995, p. 346).

Although post-structuralist ideas have had a highly problematic emergence within class analysis, they have been more readily adopted within gender studies, although even here their acceptance has not been complete, not least because they may be taken to imply there is no essential basis for gender differences. An increasing number of geographers have seen social identities as socially and culturally created and have sought to examine the processes by which these identities are created and their connections with rural social life and discourses of rurality. At present, the largest quantity of this work lies in the area of gender identities and relations, examined by Little (1987), Nead (1988), Brandth (1995), Little and Austin (1996) and Agg and Phillips (1997) among others. In addition, other forms of identity such as those related to sexuality and race are also beginning to be analysed by rural researchers.

As well as promoting several new research foci within rural social geography, the entrance of post-modern and post-structural ideas into rural geography has also impacted on the study of some rather longer-recognized aspects of social geography. A number of researchers (e.g. Gorton *et al.*, 1994; Phillips, 1993) have questioned the degree to which concepts such as counterurbanization and gentrification refer to unitary phenomena. In addition, the study of social groups, such as the disabled, has been transformed from being a somewhat marginalized subject of research into being very much part of a mainstream study of marginalized or neglected others (cf. Gant, 1991; Gant and Smith, 1988; Philo, 1992b). Similarly, there has been a re-examination of rural recreation and leisure (e.g. Clarke *et al.*, 1994; McNaughton, 1995; Urry, 1992; 1995a; 1995c), rural resource planning and conflict (Burgess, 1992; Harrison and Burgess, 1994; Whatmore and Boucher, 1993) and rural deprivation (Cloke, 1995; Cloke *et al.*, 1994; Goodwin *et al.*, 1995; Lawrence, 1995, p. 297), using notions of discourse and multiple, fragmented and neglected subjects.

The critics of post-modernism and post-structuralism

Although post-modernism and post-structuralism have made a considerable impact on rural social geography within a relatively short period, several people have raised important objections and questions about the adoption of these philosophies. Philo's article, 'Neglected rural others' stimulated two articles by Murdoch and Pratt (1993; 1994) which argued, among other things, that giving voice to neglected others is insufficient both theoretically and practically. In discussing theoretical insufficiency, they suggest that giving voice to others does not address the causes which lead to marginalization and neglect; and as for the practice of rural life, they suggest that Philo's approach will leave him at the margins of the exercise of power and his calls for 'taking difference into account' will go unheeded. By contrast, they call for a brand of rural studies which can

> reveal the ways of the 'powerful' exploring the means by which they make and sustain 'their' domination (perhaps in the hope that such knowledge could become a 'reservoir' to be drawn upon by oppositional actors)?. (Murdoch and Pratt, 1994, p. 85)

It should also

> influence the decisions of the 'powerful' such as policy-makers in the hope that they might be persuaded to produce more effective and just interventions in the world. (Murdoch and Pratt, 1994, p. 85)

In a similar vein, Phillips (1994) has drawn upon the work of the German philosopher Jürgen Habermas to argue that rural researchers need to recognize 'the difference that their knowledge can make' and to become involved in

> the project of establishing 'procedures of discursive will-formation that would put participants themselves in positions to realize concrete possibilities for better and less threatened life, on their own initiative and in accordance with their own needs and insights'. (Phillips, 1994, p. 118; quote from Habermas, 1989, p. 69)

One of the arguments underlying this project is that language does not form and operate entirely within a linguistic and mental realm, but that it is frequently used within practical contexts. In Habermas' terms, language has both 'world generative' aspects – it is through language that we come to know and define a meaningful world to us – and it has 'problem solving capacities' – it 'functions as a medium for dealing with problems that arise within the world' (McCarthy, 1985, p. xiii). David Harvey (1992) has similarly argued that some post-modernists seem to deploy a 'discursive idealism' which sees the discursive/aesthetic as the only determinant of social life in the sense that nothing exists outside representations. By contrast, Harvey argues there is 'a non-fantasy world . . . against which we can get some truth of our understanding' and that 'we should have some commitment to representing that reality, including the differences and commonalities that exist within it . . . in order to confront the multiple and manifold injustices which exist there' (Harvey, 1992, p. 315).

There are clear objections that can be raised to the arguments of both Harvey and Habermas; see for example Deutsche (1991) and Phillips (1994) respectively. They do, however, point to the need for rural social geographers to consider in more detail such seemingly esoteric subjects as the constitution of the subject and the formation of language. This would suggest there is likely to be a series of further movements of theory into rural social geography. Both Harvey and Habermas also express high degrees of allegiance to the value of a broadly Marxist political economy perspective. There is clearly, therefore, a continuing and as yet very much unresolved tension between the current streams of thought reconfiguring the social geography of the rural.

CONCLUSION

This chapter has sought to outline some of the differing social theories and philosophical arguments which have been circulating within rural social geography in recent years. It has been suggested that since the early 1980s there have been two major reconfigurations of thought within rural social geography. First, a number of rural geographers began to argue for a 'radical' or 'critical' approach to the study of rural social life, which quickly became translated into a series of broadly Marxist forms of political economic analysis. Second, and more recently, the study of rural social life has been influenced by ideas of 'difference', 'discourse' and 'the subject', stemming from post-modernist and post-structuralist cultural and social theory.

These two reconfigurations have transformed rural social geography radically from its earlier emphases on the enumeration of population change, resource conflicts and their management, access of social groups to services and changes in rural community life. It is possible to see these reconfigurations as breaking down restrictions placed on the rural social geographical imagination by empiricist and logical positivist approaches; for a fuller exposition of this claim see Phillips (1997). This has made contemporary rural social geography a more diverse, and arguably a more exciting, area of study than it has ever been. It has probably made it more intellectually, politically and ethically challenging as well. As Philo (1997) has put it, rural geography is no longer, if indeed it ever was, 'simple studies of fixed unproblematic and supposedly well-known people and places' and 'peaceful pastures free from chaos and confusion, passion and politics'. Long may it remain so.

REFERENCES

Agg, J. and Phillips, M. (1997) Neglected gender dimensions of rural social restructuring. In Boyle, M. and Halfacree, K. (eds) *Migration into rural areas: theories and issues.* Wiley, London.

Barlow, J. (1986) Landowners, property owners and the rural locality. *International Journal of Urban and Regional research,* **10**, 309–29.

Baudrillard, J. (1983a) *In the shadow of the silent majorities.* Semiotext(e), New York.

Baudrillard, J. (1983b) *Simulations.* Semiotext(e), New York.

Bell, D. and Valentine, G. (1995) Queer country: rural lesbian and gay lives. *Journal of Rural Studies,* **11**(2), 113–22.

Bell, M. (1994) *Childerley: nature and morality in a country village.* Chicago University Press, Chicago.

Berry, B. (1976) The counterurbanization process: urban America since 1970. In Berry, B. (ed) *Urbanization and counterurbanization.* Sage, Beverley Hills CA.

Bolton, N. and Chalkley, B. (1990) The population turnaround: a case study of North Devon. *Journal of Rural Studies,* **6**(1), 29–43.

Bourdieu, P. (1984) *Distinction: a social critique of the judgement of taste.* Routledge, London.

Bradley, T. (1981) Capitalism and the countryside: rural sociology as political economy. *International Journal of Urban and Regional research,* **5**, 581–87.

Brandth, B. (1995) Rural masculinity in transition: gender images in tractor advertisements. *Journal of Rural Studies,* **11**(2), 123–33.

Brown, D. and Wardwell, J. (1980) *New directions in urban–rural migration.* Academic Press, New York.

Burgess, J. (1992) The cultural politics of economic development and nature conservation. In Anderson, K. and Gale, F. (eds) *Inventing places: studies in cultural geography.* Longman Cheshire, Melbourne, pp. 235–51.

Buttel, F. and Newby, H. (1980) (eds) *The rural sociology of advanced societies: critical perspectives.* Croom Helm, London.

Champion, A. (1989) *Counterurbanization: the changing pace and nature of population deconcentration.* Edward Arnold, London.

Clark, G. (1982) *Housing and planning in the countryside.* Wiley, Chichester.

Clarke, G., Darrell, J., Grove-White, R., MacNaughten, P. and Urry, J. (1994) *Leisure landscapes.* Council for the Protection of Rural England, London.

Cloke, P. (1987) (ed) *Rural planning: policy into action?* Harper and Row, London.

Cloke, P. (1989a) Rural geography and political economy. In Peet, R. and Thrift, N. (eds) *New models in geography: the political economy perspective,* vol 1. Unwin Hyman, London, pp. 164–97.

Cloke, P. (1989b) (ed) *Rural land-use planning in developed nations.* Unwin Hyman, London.

Cloke, P. (1992) 'The countryside': development, conservation and an increasingly marketable commodity. In Cloke, P. (ed) *Policy and change in Thatcher's Britain.* Pergamon, Oxford, pp. 269–95.

Cloke, P. (1993a) The countryside as commodity: new rural spaces for leisure. In Glyptis, S. (ed) *Leisure and the environment*. Belhaven, London, pp. 53–67.

Cloke, P. (1993b) On 'problems and solutions'. The reproduction of problems for rural communties in Britain during the 1980s. *Journal of Rural Studies*, **9**(2), 113–21.

Cloke, P. (1994) (En) culturing political economy: a life in the day of a 'rural geographer'. In Cloke, P., Doel, M., Matless, D., Phillips, M. and Thrift, N. *Writing the rural: five cultural geographies*. Paul Chapman, London, pp. 149–90.

Cloke, P. (1995) Research and rural planning: from Howard Bracey to discourse analysis. In Cliff, A.D., Gould, P.R., Hoare, A.G. and Thrift, N.J. (eds) *Diffusing geography: essays for Peter Haggett*. Blackwell, Oxford, pp. 112–32.

Cloke, P. and Goodwin, M. (1992) Conceptualizing countryside change: from post-Fordism to rural structured coherence, *Transactions of the Institute of British Geographers*, **17**(3), 321–36.

Cloke, P. and Little, J. (1990) *The rural state? Limits to planning in rural society*. Oxford University Press, Oxford.

Cloke, P. and Park, C. (1985) *Rural resource management*. Croom Helm, London.

Cloke, P. and Thrift, N. (1987) Intra-class conflict in rural areas. *Journal of Rural Studies*, **3**, 321–33.

Cloke, P. and Thrift, N. (1990) Class change and conflict in rural areas. In Marsden, T., Lowe, P. and Whatmore, S. (eds) *Rural restructuring*. David Fulton, London, pp. 165–81.

Cloke, P., Phillips, M. and Rankin, R. (1991) Middle-class housing choice: channels of entry into Gower, South Wales. In Champion, T. and Watkins, C. (eds) *People in the countryside: studies of social change in rural Britain*. Paul Chapman, London, pp. 38–51.

Cloke, P., Milbourne, P. and Thomas, C. (1994) *Lifestyles in rural England*. Rural Development Commission, London.

Cloke, P., Phillips, M. and Thrift, N. (1995) The new middle classes and the social constructs of rural living. In Butler, T. and Savage, M. (eds) *Social change and the middle classes*. UCL Press, London, pp. 220–38.

Cloke, P., Phillips, M. and Thrift, N. (1997) Class, colonisation and lifestyle strategies in Gower. In Boyle, M. and Halfacree, K. (eds) *Migration to rural areas*. Wiley, London.

Connor, S. (1989) *Postmodernist culture: an introduction to theories of the contemporary*. Blackwell, Oxford.

Coppock, J. and Duffield, B. (1975) *Outdoor recreation: a spatial analysis*. Macmillan, London.

Cosgrove, S. and Daniels, S. (1988) Introduction: iconography and landscape. In Cosgrove, S. and Daniels, S. (eds) *Iconography and landscape: essays on symbolic representation, design and use of past landscapes*. Cambridge University Press, Cambridge, pp. 1–10.

D'Angelli, A. and Hart, M. (1987) Gay women, men and families in rural settings: towards the development of helping communities. *American Journal of Community Psychology*, **15**, 79–93.

Daniels, S. (1993) *Fields of vision: landscape imagery and national idenity in England and the United States*. Polity, Cambridge.

Day, G., Rees, G. and Murdoch, J. (1989) Social change, rural localities and the state: the restructuring of rural Wales. *Journal of Rural Studies*, **5**(3), 227–44.

Deutsche, R. (1991) Boys town. *Environment and Planning D*, **9**, 5–30.

Drudy, P. and Drudy, S. (1979) Population mobility and labour supply in rural regions: North Norfolk and Galway Gaelacht. *Regional Studies*, **13**, 91–99.

Fielding, A. (1982) Counterurbanisation in western Europe. *Progress in Planning*, **17**, 1–52.

Frankenberg, R. (1966) *Communities in Britain*. Penguin, Harmondsworth.

Frey, W. (1987) Migration and depopulation of the metropolis: regional restructuring or rural renaissance? *American Sociological Review*, **52**, 240–57.

Gant, R. (1991) The elderly and the disabled in rural areas: travel patterns in the north Cotswolds. In Champion, T. and Watkins, C. (eds) *People in the countryside: studies of social change in rural Britain*. Paul Chapman, London, pp. 108–24.

Gant, R. and Smith, J.A. (1988) Journey patterns of the elderly and disabled in the Cotswolds: a spatial analysis. *Social Science and Medicine*, **27**(2), 173–80.

Gilg, A. (1985) *An introduction to rural geography*. Edward Arnold, London.

Goldthorpe, J. (1982) On the service class, its formation and future. In Giddens, A. and McKenzie. G. (eds) *Social class and the division of labour*. Cambridge University Press, Cambridge, pp. 162–85.

Goldthorpe, J. and Marshall, G. (1992) The promising future of class analysis. *Sociology*, **26**, 381–400.

Goodwin, M., Cloke, P. and Milbourne, P. (1995) Regulation theory and rural research: theorising contempoary rural change. *Environment and Planning A*, **27**, 1245–60.

Gorton, M., White, J. and Chaston. (1994) Fragmentation, conflict and justice in rural localities. In *Societies in transition: conference proceedings*, vol l. Edinburgh College of Art and Heriot-Watt University, Edinburgh.

Gouldner, A. (1979) *The future of intellectuals and the rise of the new class*. Continuum, New York.

Gregory, D. (1978) The discourses of the past. *Journal of Historical Geography*, **4**(2), 161–73.

Gregory, D. (1981) Human agency and human geography. *Transactions of the Institute of British Geographers*, **6**, 1–18.

Habermas, J. (1987) *The philosophical discourse of modernity*. Polity, Cambridge.

Habermas, J. (1989) *The new conservatism: cultural criticism and the historians' debate*. Polity, Cambridge.

Halfacree, K. (1993) Locality and social representation: space, discourse and alternative definitions of the rural. *Journal of Rural Studies*, **9**, 1–15.

Halfacree, K. (1994a) Displacing the rural idyll: mobile lifestyles in a settled countryside. Paper presented at Accessing the Countryside, Rural Geography Study Group, Nottingham.

Halfacree, K. (1994b) Neo-tribes, migration and the post-productivist countryside. Paper presented at Migration Issues in Rural Areas Conference, Swansea.

Halfacree, K. (1996) Out of place in the country: travellers and the rural idyll. *Antipode*, **28**(1), 42–72.

Halliday, J. and Coombes, M. (1995) In search of counterurbanisation: some evidence from Devon on the relationship between patterns of movement and motivation. *Journal of Rural Studies*, **11**(4), 433–46.

Hamilton, P. (1990) Sociology: commentary and introduction. In Lowe, P. and Bodiguel, M. (eds) *Rural studies in Britain and France*. Belhaven, London, pp. 225–31.

Harper, S. (1989) The British rural community: an overview of perspectives. *Journal of Rural Studies*, **5**(2), 161–84.

Harrison, C. (1991) *Countryside recreation in a changing society*. TMS Partnership, London.

Harrison, C. and Burgess, J. (1994) Social constructions of nature: a case study of the conflicts over the development of Rainham Marshes. *Transactions of the Institute of British Geographers*, **19**(3), 291–310.

Harvey, D. (1992) Postmodern morality plays. *Antipode*, **24**, 300–26.

Hetherington, K. (1995) *On the homecoming of the stranger: new social movements or new sociations?* Lancaster Regionalism Group, Working Paper 39, University of Lancaster.

Hill, N. and Young, N. (1991) Support policy for rural areas in England and Wales: its assessment and qualification. *Journal of Rural Studies*, **7**(3), 191–206.

Hoggart, K. and Buller, H. (1987) *Rural development: a geographical perspective.* Croom Helm, London.

Hoggart, K., Buller, H. and Black, R. (1995) *Rural Europe: identity and change.* Edward Arnold, London.

Jessop, B. (1990) Regulation theories in retrospect and prospect. *Economy and Society*, **19**, 153–216.

Jones, O. (1995) Lay discourses of the rural: development and implications for rural studies. *Journal of Rural Studies*, **11**(1), 35–49.

Joseph, A., Keddie, P. and Smit, B. (1988) Unravelling the population turnaround in rural areas. *Canadian Geographer*, **32**, 17–39.

Joseph, A., Smit, B. and McIlravey, G. (1989) Consumer preferences for rural residences: a conjoint analysis in Ontario, Canada. *Environment and Planning A*, **21**, 47–64.

Kinsman, P. (1995) Landscape, race and national identity. *Area*, **27**(4), 300–310.

Lawrence, M. (1995) Rural homelessness: a geography without a geography. *Journal of Rural Studies*, **11**(3), 297–307.

Lawton, R. (1968) Population changes in England and Wales in the late nineteenth century: an analysis of trends by registration districts. *Transactions of the Institute of British Geographers*, **44**, 55–74.

Lewis, G. (1979) *Rural communities: a social geography.* David and Charles, Newton Abbot.

Little, J. (1987) Gender relations in rural areas: the importance of women's domestic role. *Journal of Rural Studies*, **3**(4), 335–42.

Little, J. and Austin, P. (1996) Women and the rural idyll. *Journal of Rural Studies*, **12**, 101–111.

Lovering, J. (1989) The restructuring debate. In Peet, R. and Thrift, N. (eds) *New models in geography: the political economy perspective*, vol 1. Unwin Hyman, London, pp. 198–223.

Lowe, P., Murdoch, J., Marsden, T., Munton, R. and Flynn, A. (1993) Regulating the new rural spaces: the uneven development of land. *Journal of Rural Studies*, **9**(3), 205–22.

McCarthy, K. and Morrison, P. (1977) The changing demographic and economic structure of non-metropolitan areas. *International Regional Science Review*, **2**, 123–42.

McCarthy, T. (1985) Introduction. In Habermas, J. *The philosophical discourse of modernity.* Cambridge University Press, Cambridge, pp. vii–xvii.

McCleary, A. (1991) Population and social conditions in remote areas: the changing character of the Scottish Highlands and Islands. In Champion, T. and Watkins, C. (eds) *People in the countryside: studies of social change in Britain.* Paul Chapman, London, pp. 144–59.

McNaughton, P. (1995) Public attitudes to countryside leisure: a case study of ambivalence. *Journal of Rural Studies*, **11**(2), 135–47.

Marsden, T. (1995) Rural change, regulation theory and sustainability. *Environment and Planning A*, **27**, 1180–92.

Marsden, T. and Murdoch, J. (1990) *Restructuring rurality: key areas for development in assessing rural change.* ESRC Countryside Change Initiative, Working Paper 4.

Marsden, T., Murdoch, J., Lowe, P., Munton, R. and Flynn, A. (1993) *Constructing the countryside.* UCL Press, London.

Matless, D. (1994) Doing the English Village, 1945–90: an essay in imaginative geography. In Cloke, P., Doel, M., Matless, D., Phillips, M. and Thrift, N. *Writing the rural: five cultural geographies.* Paul Chapman, London, pp. 7–88.

Miller, S. (1996) Class, power and social construction: issues of theory and application in thirty years of rural studies. *Sociologica Ruralis*, **36**, 93–116.

Mohan, G. (1994) Destruction of the con: geography and the commodification of knowledge. *Area*, **26**(4), 387–90.

Mormont, (1987) Rural nature and urban nature. *Sociological Ruralis*, **28**, 3–21.

Moseley, M. (1979) *Accessibility: the rural challenge*. Methuen, London.

Moses, A. and Buckner, J. (1980) Special problems of rural gay clients. *Human Services in the Rural Environment*, **5**, 22–7.

Munton, R. (1995) Regulating rural change: property rights, economy and environment – a case study from Cumbria, UK. *Journal of Rural Studies*, **11**(3), 269–84.

Murdoch, J. (1995a) Actor-networks and the evolution of economic forms: combining description and explanation in theories of regulation, flexible specialisation and networks. *Environment and Planning A*, **27**, 731–57.

Murdoch, J. (1995b) Middle class territory? Some remarks on the use of class analysis in rural studies. *Environment and Planning A*, **27**, 1213–30.

Murdoch, J. and Marsden, T. (1991) *Reconstituting the rural in an urban region: new villages for old?* Countryside Change Working Paper, University of Newcastle.

Murdoch, J. and Marsden, T. (1994) *Reconstituting rurality: class, community and power in the development process*. UCL Press, London.

Murdoch, J. and Marsden, T. (1995) The spatialization of politics: local and national actor spaces in environmental conflict. *Transactions of the Institute of British Geographers*, **20**(3), 368–81.

Murdoch, J. and Pratt, A. (1993) Rural studies: modernism, postmodernism and the 'post rural'. *Journal of Rural Studies*, **9**(4), 411–27.

Murdoch, J. and Pratt, A. (1994) Rural studies of power and the power of rural studies: a reply to Philo. *Journal of Rural Studies*, **10**(1), 83–87.

Nead, L. (1988) *Myths of sexuality: representations of women in Victorian Britain*. Blackwell, Oxford.

Newby, H. (1986) Locality and rurality: the restructuring of rural social relations. *Regional Studies*, **20**, 209–16.

Newby, H. (1987) *Country life: a social history of rural England*. Weidenfeld and Nicholson, London.

Newby, H. and Buttel, F. (1980) Towards a critical rural sociology. In Buttel, F. and Newby, H. (eds) *The rural sociology of advanced societies: critical perspectives*. Croom Helm, London, pp. 1–35.

Newby, H., Bell, C., Rose, D. and Saunders, P. (1978) *Property, paternalism and power: class and control in rural England*. Hutchinson, London.

Pahl, R. (1965) Class and community in English community villages. *Sociologica Ruralis*, **5**, 5–23.

Pahl, R. (1966) The rural–urban continuum. *Sociologica Ruralis*, **6**, 299–327.

Pahl, R. (1989) Is the emperor naked? Some questions of the adequacy of sociological theory in urban and regional research. *International Journal of Urban and Regional Research*, **13**, 711–20.

Parker, K. (1984) *A tale of two villages: the story of the Integrated Rural Development Experiment in the Peak District, 1981–84*. Peak Park Joint Planning Board, Bakewell.

Patmore, J. (1983) *Recreation and resources*. Blackwell, Oxford.

Pettigrew, P. (1987) A bias for action: industrial development in mid-Wales. In Cloke, P. (ed) *Rural planning: policy into action?* Harper and Row, London, pp. 102–21.

Phillips, D. and Williams, A. (1982) *Rural housing and the public sector*. Gower, Farnborough.

Phillips, M. (1993) Rural gentrification and the processes of class colonisation. *Journal of Rural Studies*, **9**(2), 123–40.

Phillips, M. (1994) Habermas, rural studies and critical social theory. In Cloke, P., Doel, M., Matless, D., Phillips, M. and Thrift, N. *Writing the rural: five cultural geographies*. Paul Chapman, London, pp. 89–126.

Phillips, M. (1997) The restructuring of social imaginations in rural geography. *Journal of Rural Studies*.

Phillips, M. (forthcoming-a) Fragmentation and contestation within the British rural middle classes. *Journal of Rural Studies*.

Philo, C. (1992a) Introduction, acknowledgements and brief thoughts on older words and older worlds. In Philo, C. (ed) *New words, new worlds: reconceptualising social and cultural geography*. Social and Cultural Geography Study Group, Lampeter, pp. 1–13.

Philo, C. (1992b) Neglected rural geographies: a review. *Journal of Rural Studies*, **8**(2), 193–207.

Philo, C. (1993) Postmodern rural geography? A reply to Murdoch and Pratt. *Journal of Rural Studies*, **9**(4), 429–36.

Philo, C. (1997) Review: writing the rural. *Environment and Planning A*, **29**, 943–50.

Pratt, G. (1994) Poststructuralism. In Johnson, R.J., Gregory, D. and Smith, D. (eds) *The dictionary of human geography*, 3rd edn. Blackwell, Oxford, pp. 468–69.

Redfield, R. (1947) The folk society. *American Journal of Sociology*, **52**, 293–308.

Rees, G. (1984) Rural regions in national and international economies. In Bradley, T. and Lowe, P. (eds) *Locality and rurality*. GeoBooks, Norwich, pp. 27–44.

Robinson, G. (1990) *Conflict and change in the countryside*. Belhaven, London.

Rounds, K. (1988) AIDS in rural areas: Challenges to providing care. *Social Work*, May–June, 257–61.

Savage, M. (1994) Class analysis and its futures. *Sociological Review*, **42**, 531–48.

Savage, M. and Butler, T. (1995) Assets and the middle classes in contemporary Britain. In Savage, M. and Butler, T. (eds) *Social change and the middle classes*. UCL Press, London, pp. 345–57.

Savage, M., Dickens, P. and Fielding, A.J. (1988) Some social and political implications of the contemporary fragmentation of the service class. *International Journal of Urban and Regional Research*, **12**, 455–76.

Savage, M., Barlow, J., Dickens, P. and Fielding, T. (1992) *Property, bureaucracy and culture: middle class formation in contemporary Britain*. Routledge, London.

Shields, R. (1991) *Places on the margins: alternative geographies of modernity*. Routledge, London.

Shields, R. (1992) Spaces for the subject of consumption. In Shields, R. (ed) *Lifestyle shopping: the subject of consumption*. Routledge, London, pp. 1–20.

Short, J.R. (1991) *Imagined country: society, culture and environment*. Routledge, London.

Shucksmith, M. (1981) *No homes for locals?* Gower, Farnborough.

Sibley, D. (1995) *Geographies of exclusion: society and difference in the West*. Routledge, London.

Strachan, A. (1988) Business development in the rural south Midlands. *Midland Geographer*, **11**, 13–21.

Thrift, N. (1987) Manufacturing rural geography. *Journal of Rural Studies*. **3**, 77–81.

Thrift, N. (1989) Images of social change. In Hamnett, C., McDowell, L. and Sarre, P. (eds) *The changing social structure*. Sage, London, pp. 12–42.

Tickell, A. and Peck, J. (1992) Accumulation, regulation and the geographies of post-Fordism. *Progress in Human Geography*, **16**, 190–218.

Tönnies, F. (1957) *Community and association*. Routledge & Kegan Paul, London.

Urry, J. (1984) Capitalist restructuring, recomposition and the regions. In Bradley, T. and Lowe, P. (eds) *Locality and rurality: economy and society in rural regions*. GeoBooks, Norwich, pp. 45–64.

Urry, J. (1990) *The tourist gaze: leisure and travel in contemporary societies*. Sage, London.

Urry, J. (1992) The tourist gaze and the environment. *Theory, Culture and Society*, **9**(3), 1–26.

Urry, J. (1995a) The making of the Lake District. In Urry, J. *Consuming places*. Routledge, London, pp. 193–210.

Urry, J. (1995b) A middle-class countryside? In Butler, T. and Savage, M. (eds) *Social change and the middle classes*. Routledge, London, pp. 205–19.

Urry, J. (1995c) Social identity, leisure and the countryside. In Urry, J. *Consuming places*. Routledge, London, pp. 211–29.

Ward, D. (1995) Darcy's wet shirt draws visitors, *The Guardian*, 7 November, p. 10.

Whatmore, S. and Boucher, S. (1993) Bargaining with nature: the discourse and practice of environmental planning gain. *Transactions of the Institute of British Geographers*, **18**, 166–78.

Whatmore, S., Munton, R. and Marsden, T. (1992) The rural restructuring process: emerging divisions of agricultural property rights. *Regional Studies*, **24**(3), 235–45.

Williams, R. (1985) *The country and the city*. Hogarth, London.

Wilson, A. (1992) *The culture of nature: North American landscape from Disney to Exxon Valdez*. Blackwell, Oxford.

Winter, M. (1984) Agrarian class structures and family farming. In Bradley, T. and Lowe, P. (eds) *Locality and rurality: economy and society in rural regions*. GeoBooks, Norwich, pp. 113–28.

Wirth, L. (1938) Urbanism as a way of life. *American Journal of Sociology*, **44**, 1–24.

Woodruffe, B. (1976) *Rural settlement policies and plans*. Oxford University Press, Oxford.

Wright, E.O. (1978) *Class crisis and the state*. New Left Books, London.

Wright, E.O. (1979) *Class structure and income determination*. Academic Press, New York.

Wright, E.O. (1985) *Classes*. Verso, London.

Wynne, D. (1990) Leisure, lifestyle and the social construction of social position. *Leisure Studies*, **9**, 21–34.

PART 2

CHANGES IN THE EXTENSIVE USE OF RURAL LAND

CHAPTER 4

FROM AGRICULTURAL PRODUCTIVISM TO POST-PRODUCTIVISM

Brian Ilbery and Ian Bowler

INTRODUCTION

Agriculture in developed market economies has undergone a substantial restructuring in the postwar period (Marsden *et al.*, 1986) and two major phases of change can be identified. The *productivist* phase, where the emphasis was placed on raising farm output, lasted from the early 1950s to the mid-1980s and was characterized by a continuous modernization and industrialization of agriculture. The *post-productivist* phase, termed the *post-productivist transition* (PPT) by a number of researchers (e.g. Lowe *et al.*, 1993), where the aim is to reduce farm output, is now a decade old and characterized by the integration of agriculture within broader rural economic and environmental objectives (Bowler and Ilbery, 1993; Shucksmith, 1993). Both phases have been manipulated by state intervention in agriculture; indeed the PPT in Europe and the United States is being shaped by a rapidly changing international policy context, especially the reforms of the Common Agricultural Policy (CAP) effective from 1992; the GATT agreement on world agricultural trade, effective from 1993; and the Agenda 21 declaration from the Earth Summit at Rio de Janeiro in 1992.

Note that productivist farm systems have not been replaced by post-productivist systems: the two diverging pathways coexist. Thus intensive, high input–high output farming, with an emphasis on food *quantity*, is still being encouraged. But this is now complemented by the development of low input–low output farming, with an emphasis on sustainable farming systems and food *quality*. These divergent pathways are likely to become more spatially differentiated, at regional and national scales, as developed market economies progress from one *food regime* to another (see below). Indeed, the uneven development of the productivist regime seems likely to be deepened during the PPT.

This chapter is structured into five main sections. The first section outlines three main theoretical explanations of agricultural productivism, and the second section provides empirical evidence of the development of productivist agriculture in the European Union (EU) and the United States. The third and fourth sections are respectively concerned with theoretical explanations and empirical evidence of the PPT. The fifth and final section describes prospects for the development of a 'sustainable' agriculture.

COMPETING THEORIZATIONS ON AGRICULTURAL PRODUCTIVISM

Several theoretical conceptualizations help to explain the dynamics of agriculture in developed market economies during the productivist phase. Three main schools of thought can be identified: commercialization, commoditization, and industrialization (Bowler, 1992a, pp. 19–27).

Commercialization

Emerging from modernization theory, with strong neoclassical underpinnings, this approach emphasizes the importance of economic factors in agricultural change and suggests that traditional, family labour farms are transformed by the introduction of supply–demand relations in a market economy (Vandergeest, 1988). The degree of commercialization is measured by the proportion of farm produce sold in the market. Agricultural commercialization is the basis of economic development, and the integration of farm households into the rural economy and society is an integral part of the process.

Three main elements of commercialization in agriculture are important. First, the process is strongly supported by research and development which produces new farm technologies, industries to manufacture the necessary inputs, and educational programmes to provide farmers with the necessary skills to apply those new inputs. Technology is thus a cause (not an effect) of agricultural modernization and economic development. Moreover, the state is deeply implicated in the type of technology that is created and communicated to the farm sector through the funding of research institutes, experimental farms, private sector research and development, and the farm extension (advisory) services. The result is the substitution of labour by new farm technology; for example, the purchase of such modern inputs as machinery, seeds, chemicals and fertilizers (i.e. the capitalization of agriculture).

Secondly, commercialization leads to a dual farming economy, with a mixture of 'traditional' family labour farms and 'modern' capitalistic farms. The capitalistic farms are assumed to be more technically efficient and more able to respond to the changing demands of the market than the 'traditional' family labour farm sector. Family farms tend to be less capital-intensive and survive on high quality work and enterprises which yield few scale economies. Indeed, the decline in farm numbers is explained by the failure of some farm households to adopt modern farm technology. Such economic concepts as maximization of profits, economic efficiency and competitiveness are used to explain the increasing integration of the farm sector into other parts of the food supply system (e.g. with food processors and retailers).

Thirdly, commercialization theory uses distance from major urban-industrial concentrations, regional farm-size structures, and the distribution of high quality natural resources to help explain why agricultural modernization occurs earlier and to a greater extent in some areas than in others. Thus urban fringes and regions with high-quality soils are the first to be transformed by the processes of commercialization, before the new organizational forms and technology 'diffuse' to other areas.

Commoditization

The commoditization approach emerged from dependency theory in the 1960s and 'locates economic analysis within specific social formations and explains the development processes in terms of the benefits and costs they carry for different social classes' (Redclift, 1984, p. 5). Emphasis is thus placed on social rather than economic structures and relations, and many advocates of commoditization theory adopt a political economy (structuralist) approach to their work. It is contended that farm households become dependent on goods obtained in the market and are therefore drawn into commercial exchanges in order to acquire income for the purchase of necessary farm inputs (Vandergeest, 1988). The commoditization approach, unlike commercialization, places emphasis on farm inputs rather than farm outputs sold in the market.

Several points follow from this approach. First, farming structures are interpreted as class structures, with the polarization between 'traditional' and 'modern' farms being a result of the capitalist accumulation process. The capitalist landowning class occupies or owns large-scale wage-labour farms, whereas the marginalized and ultimately landless class supply wage-labour for the modern capitalist farms. In relation to the latter, Friedmann (1986) used the term *simple (or petty) commodity production* to describe small farms based solely on family labour (family labour farms). Nevertheless, this evolutionist theory, which predicts the complete elimination of family labour farms, is flawed because of the continued survival of such farms. For example, many large-scale, modern farm businesses are owned and operated exclusively by family labour (Marsden and Symes, 1984), whereas many family labour farms use hired labour at different stages in their development.

Secondly, control by external capital is progressively exercised over the family labour farm through a process of *subsumption* and the extraction of surplus value. Subsumption can be either *real* or *formal*. Although real subsumption is achieved through the direct ownership of the means of production (farmland) by external capitals (i.e. vertical integration by agri-industrial complexes), formal subsumption implies legal ownership remaining with the family labour farm but effective control passing to external capitals, often through a system of agricultural contracts. Indeed, the processing of farm products is increasingly controlled by food processing industries which prefer to obtain their raw materials through a system of forward contracts with farmers rather than by the direct ownership of land (Hart, 1992).

Thirdly, the uneven transformation of agriculture is a function of the penetration of particular farming systems (and thus farming regions) by external capitals. External capitals (often in the form of transnational corporations) penetrate those areas where greatest financial returns can be obtained, and they are prepared to switch their investments between regions when profits fall. Marsden *et al.* (1987) refer to this process as 'waves of capitalization' and space is seen as either an opportunity or constraint in the process of commoditization.

It has been difficult to provide consistent empirical support for the theoretical ideas advocated by the commoditization school, especially because of the persistence of the family labour farm under advanced capitalism. Indeed, farm households have a greater degree of autonomy than theorized and there is a need to incorporate behavioural attributes (human agency) into political economy perspectives on agriculture. The family labour farm has survived and, in some cases, prospered by adaptation to capitalism through different 'developmental pathways' (see below). A major example of this is the generation of income from non-agricultural enterprises on and/or off the farm by members of the farm household, a process called *pluriactivity*.

Industrialization

Drawing upon concepts from both the commercialization and commoditization schools, the industrialization approach adopts the food-supply system as its organizing framework and focuses on long-run changes in capitalist agriculture in response to biophysical and natural production processes. The uniqueness of agriculture in terms of such processes prevents the unified transformation of agriculture by industrial capitals (Goodman *et al.*, 1987). Rather, progress towards industrialization has occurred in a series of steps. First, machinery replaced animal power; secondly, the introduction of such inputs as hybrid seeds, fertilizers and agrochemicals allowed the modification or replacement of natural biological processes; and thirdly, industrial substitutes were developed for agricultural products, such as sweeteners for sugar and nylon for cotton. As Goodman and Redclift (1991, pp. 201–2) emphasize, 'some processes were taken out of farm production (appropriation) and others were substituted, especially in food processing activities (substitution)'.

Two theoretical concepts thus dominate the industrialization approach:

1. *Appropriationism* is taken from the commoditization literature and is defined as the 'discontinuous but persistent transformation into industrial activities of certain parts of the agricultural production process, and their subsequent reintroduction in the form of purchased farm *inputs*'. Good examples would be tractors replacing animal power and synthetic chemicals replacing manure. The appropriations are (i) *partial*, as not all processes are affected; (ii) *discrete*, as specific parts of agricultural production have been appropriated; and (iii) *discontinuous*, in that innovatory phases are introduced from time to time.
2. *Substitutionism* focuses on *outputs* rather than inputs, is concerned with the increased utilization of non-agricultural raw materials and the creation of industrial substitutes for food and fibre. Le Heron (1993, p. 37) remarks, 'discrete elements of production processes have been taken over by industry and products of agriculture have been substituted by outputs from a growing complex of food and fibre industries'. Agriculture is thus a source of raw materials for mass industrial processing and marketing and, where possible, production from farms is not used at all.

The industrialization of agriculture has become increasingly global, with recent developments in biotechnology being international in nature (Le Heron, 1993). Large integrated agribusiness corporations are seeking out low-cost agricultural raw materials on the international market for their inputs. Thus the spatial dimension is now a global one. The rate and direction of appropriationism and substitutionism are largely determined by prevailing levels of technological change, but in addition there are close relationships with state intervention in agriculture. Indeed, Newby (1982, p. 138) has observed how 'the technological transformation of agriculture can be viewed no longer as a product of the hidden hand of the market, rather it is the outcome of deliberate policy decisions by the state, consciously pursued and publicly encouraged'. Goodman and Redclift (1991) also describe a technology–policy model, which started with the New Deal farm support programmes in the United States during the 1930s and later spread to other industrial economies. Such support policies encouraged farm-based accumulation and helped to link farmers more closely to 'upstream' and 'downstream' industries.

Unlike the first two schools of thought, the industrialization school highlights the increasing role of the state in influencing the trajectory of agricultural change. Indeed, the interaction of cheap food policies and industrial capital has exerted increasing control over agricultural production, processing and manufacture. However, since the mid-1980s there has been a reappraisal of national state policies for agriculture, with a reduction in price protection and subsidies. Negotiations now often take place between states and groups of states, at the international level.

In fact, Goodman and Redclift (1989) argue that the relations between agriculture and industry have historically been more global than generally assumed. They used the concept of *food regimes* to link international relations of food production and consumption to forms of accumulation and regulation under capitalist systems from the 1870s onwards. From this perspective, three major food regimes can be identified (Table 4.1): the first involves settler colonies supplying unprocessed and semiprocessed foods and materials to the metropolitan core of North America and

Table 4.1   Food regimes from the 1870s onwards

Character	First regime	Second regime	Third regime
Products	Grain, meat	Grain, meat, durable food	Fresh, organic, reconstituted
Period	1870 to WWI[a]	1920s to 1980s	1990s
Capital	Extensive	Intensive	Flexible
Food systems	Exports from family farms in settler colonies	Transnational restructuring of agriculture to supply mass market	Global restructuring, with financial circuits linking production and consumption
Characteristics	Culmination of colonial organization of precapitalist regimes Rise of nation states	Decolonization Consumerism Growth of forward and backward linkages from agriculture	Globalization of production and consumption Disintegration of national agrofood capital and state regulation Green consumers

[a] WWI = World War I.
Source: Adapted from Le Heron (1992).

western Europe; the second regime relates to the productivist phase of agricultural change, focused on North America and the development of agro-industrial complexes and grain-fed livestock; and the third regime involves greater flexibility, including the production of fresh fruit and vegetables for the global market, the continued reconstitution of food through industrial and bioindustrial processes, and the supply of inputs for 'elite' consumption in the north.

EMPIRICAL EVIDENCE OF PRODUCTIVIST AGRICULTURE

This section turns to the empirical detail of change in the productivist era and emphasizes the differentiation, or uneven development, of agriculture. The discussion begins by simplifying the complexity of agricultural change into three structural dimensions: intensification, concentration and specialization (Bowler, 1985a). Each dimension comprises a set of primary process responses with accompanying secondary process responses (Table 4.2).

Three structural dimensions in productivist agriculture

Intensification can be measured either by increased farm inputs (e.g. capital, fertilizer, agrochemicals) or farm outputs (e.g. production of cereals, meat, livestock products) per hectare of agricultural land. For example, in the four decades leading to the mid-1980s all of the countries in the EU were characterized, to varying degrees, by intensification (Jansen and Hetsen, 1991), a process already described as the 'capitalization' of agriculture. This feature is illustrated in Table 4.3 for inputs of inorganic fertilizers within the EU during the productivist era: inputs increased threefold in France between 1956 and 1985, and nearly doubled in the Netherlands over the same period.

The trend towards *concentration* in agriculture describes the increasing proportion of total productive resources (especially labour and capital) or farm production (outputs) located in a smaller number of census units, such as parishes or counties. Looking at outputs, for example, Bowler (1985a) has shown how the production of wheat, potatoes, milk and oilseeds within the EU became increasingly concentrated in Denmark, Ireland, the United Kingdom and West Germany between 1969 and 1981, whereas the production of fresh fruit, pigmeat, eggs and sheepmeat became more concentrated in France, Belgium, the Netherlands and Italy.

Increased concentration is also evident in the distribution of *land* at the farm level. Economies of scale on larger farms allow fixed costs to be spread over greater volumes of production, reducing production costs per unit of output. The process of capital accumulation and concentration in the competitive enlargement of farm units has produced a sustained fall in the number of smaller farms and the concentration of land into larger farms. The number of people employed in agriculture has also been falling; the annual average fall is 3% in the EU since 1970 and values for individual countries are given in Table 4.4.

Table 4.2   The industrialization of agriculture

Primary process responses	
Structural dimension	Outcome
Intensification	Purchased inputs (capital) replace labour and substitute for land, increasing dependence on agro-inputs industries Mechanization and automation of production processes Application of developments in biotechnology
Concentration	Fewer but larger farming units Production of most crops and livestock concentrated on fewer farms, regions and countries Sale of farm produce to food processing industries – increasing dependence on contract farming
Specialization	Labour specialization, including the management function Fewer farm products from each farm, region and country
Secondary consequences	
Structural dimension	Outcome
Intensification	Development of supply (requisites) cooperatives Rising agricultural indebtedness Increasing energy intensity and dependence on fossil fuels Overproduction for the domestic market Destruction of environment and agro-ecosystems
Concentration	Development of marketing cooperatives New social relations in rural communities Inability of young to enter farming Polarization of the farm size structure Corporate ownership of land Increasing inequalities in farm incomes between farm sizes, types and locations State agricultural policies favouring large farms and certain regions
Specialization	Food consumed outside region where it was produced Increased risk of system failure Changing composition of the workforce Structural rigidity in farm production

Source: Bowler (1985a).

Table 4.3   Application of inorganic fertilizers in countries of the EU[a]

Year	West Germany	France	Netherlands	United Kingdom
1956	148 (2 114)	56 (1 924)	201 (468)	– (–)
1965	209 (2 897)	93 (3 123)	250 (566)	79 (1 555)
1975	251 (3 300)	152 (4 850)	306 (638)	95 (1 800)
1985	265 (3 185)	181 (5 694)	346 (701)	135 (2 524)

[a] The fertilizers are nitrogen, phosphate and potash; the figures are kg ha^{-1} and in parentheses thousands of tonnes.
Source: Compiled from data in Commission of the European Communities (various years).

Table 4.4   Agriculture in the economies of EU member states circa 1990

Country	A Employment (%)	B GDP (%)	C $B/A \times 100$
Belgium	2.7	2.1	78
Denmark	5.5	3.7	67
Germany	3.3	1.5	45
Greece	21.6	13.9	64
Spain	10.7	4.6	43
France	5.8	3.3	57
Ireland	13.8	9.5	69
Italy	8.5	3.6	42
Luxembourg	3.1	2.2	71
Netherlands	4.5	4.0	89
Portugal	17.5	5.3	30
United Kingdom	2.2	1.4	64
EU (12)	6.2	2.9	47

GDP = gross domestic product.

Source: Compiled from data in Commission of the European Communities (various years).

Specialization in agriculture is based on the economies of scale that can be gained by limiting production to a few products in a farm business. And in an increasingly sophisticated technological environment, there are management advantages in learning and applying specialist skills and knowledge. Thus specialization can be observed in the functions of the labour force, the types of farm equipment employed, and the resulting land use. When aggregated for groups of farms, increasingly specialized and differentiated agricultural regions can be identified (Bowler and Ilbery, 1989). Figure 4.1 shows the regional pattern of specialization in land use within the EU for 1987, using Hr as a measure of entropy between five major land uses (cereals, grassland, industrial crops, fruit/vegetables and vines).

These dimensions of productivist agriculture, when taken together, have produced an 'industrialization' of agriculture (Troughton, 1986), a developmental trend likely to be perpetuated unevenly on farms and in farming regions by two recent developments in biotechnology. First, developments in genetic engineering offer considerable potential for raising yields: new drought and pest-resistant crop varieties; grass and crop varieties requiring lower levels of artificial fertilizer to achieve equal yields; and livestock maturing at earlier ages or providing higher yields of milk or meat. Such developments in genetic engineering for agriculture pose similar unresolved ethical problems as in the field of human medical science. Secondly, the agrofood industry has developed new technologies for enzymes, biosensors and fermentation (Traill, 1989). For example, developments in biotechnology (already described as appropriationism and substitutionism) have led to textured vegetable protein becoming a competitor for meat products, isoglucose becoming a competitor for sugar, and ethanol becoming a petrochemical substitute. To feed these new processes, agrofood firms are demanding crops with improved processing characteristics,

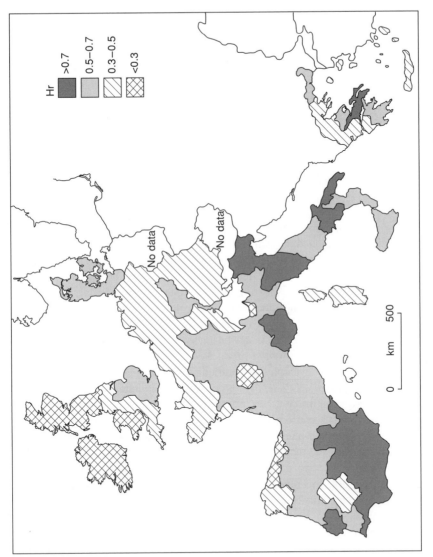

Figure 4.1 Regional patterns of agricultural specialization in the European Union, 1987: see text for explanation of Hr

especially cereals, sugar and oils, and demanding greater volumes of raw material with a uniform quality. These new demands are beginning to impact on agriculture throughout the world (Le Heron, 1993).

The food production chain and external capital

Reference to the increasing importance of agrofood companies points up another major characteristic of the productivist era, namely the absorption of agriculture into the food supply system, also called the commodity chain and the agrofood complex (Whatmore, 1995). Here three types of external capital have become particularly important: agrofood companies, food retailers and the financial services sector. From a geographical perspective, the most significant feature has been the uneven integration of individual farms and farming regions into the new food supply systems (FitzSimmons, 1986). The development of large-scale farms in the EU with close economic links to external capital has been a characteristic of only certain regions, such as East Anglia, the southern Netherlands, the Paris Basin and Emilia Romagna. Farmers in other regions have been able to resist such integration into the agrofood system, with the result that productivist agriculture in the EU has become more differentiated.

Agrofood companies have developed their businesses to meet the demand for manufactured foods (canned, frozen, part-cooked and preassembled meals), and today almost all foodstuffs are subjected to some form of value-added treatment off the farm by food processors before reaching the consumer. Research by Gregor (1982) in the United States has shown how farms that are part of a large company (an agribusiness) tend to occupy locations near to large urban areas, especially in the California valleys and the Florida peninsula.

In western Europe, by contrast, agribusinesses usually avoid owning farms and instead offer renewable, forward production contracts to farmers. Three consequences for the uneven development of agriculture can be identified. First, production under contract is *spatially selective*: producers in the vicinity of the processing plants tend to be awarded the contracts. Secondly, farming under contract is *selective by farm business size*. Because processors require a constant flow of standardized raw materials through their plants, contracts tend to be placed with a small number of larger farm businesses, including farmer cooperatives. Thirdly, farming under contract is *selective by type of farming*, favouring such products as vegetables, fruit, pigs and poultry. In France, for example, 50% of poultrymeat and 93% of vegetables are marketed under contract; the figure for both products in the Netherlands is over 90%.

Since the 1950s, *food retailing* in developed countries has experienced intense restructuring. Small, independent retailers have been marginalized by large multiple supermarket chains, such as Tesco, Sainsbury and Asda in the United Kingdom (Wrigley, 1987); their numbers have been substantially reduced and they survive by offering consumers longer opening hours and specialized foods, often in locations within urban-residential areas. The large food retailers now increasingly determine prices in the market for food processors and thus for farm producers. Indeed, with the large retailers themselves prepared to search the international market for produce to put on the shelves of their stores, the global competition between sets of processors and farmers has been intensified. For example, in 1989 Argyll/Safeway (UK), Groupe Casino (France) and Koninklijke Ahold (Netherlands) formed a new joint venture company to coordinate their purchasing, distribution, marketing and information systems.

With the increasing capital needs of modern agriculture, including the rising cost of purchasing farmland, *credit and financial markets* have become important influences on agriculture, including the interest rate charged for borrowing capital, the level of farm indebtedness, and the behaviour of banks and credit institutions towards farmers when agriculture encounters financial difficulties in servicing its loans. However, few empirically based studies of the actual behaviour of the financial services sector in relation to agriculture have been published, and most research on the subject remains at a theoretical level.

The external costs of agricultural productivism

One of the most important changes to occur during the productivist era was the increased pressure on farmers to alter their behaviour towards the natural environment (nature). Farmers have always been sensitive to the conservation of the environment, although this has tended to be narrowly defined in terms of such features as the management of soil, farm woodland and moorland. But the logic of the industrialization of agriculture placed pressure on farmers to take an exploitative rather than conservative behaviour towards their natural resource base. For example, intervention by the state reduced the risks associated with agricultural specialization, and protected prices for farm products provided a financial inducement to intensify output; besides which, new farming technology was output-increasing as well as cost-reducing. In sum, financial expediency tended to override best farming practice in relation to conservation of the natural environment.

The damaging environmental effects of modern agriculture can be understood as the outcome of manipulating agro-ecosystems for financial gain. Two types of manipulation can be identified: increased energy flows through a farming system, and modification of 'natural' components within a farming system. Under *increased energy flows* can be placed the various types of pollution caused by the intensification of agriculture and the increased application of fertilizers and agrochemicals to both crop and livestock farming systems. Examples include the contamination of water resources by fertilizers leached from crop and pasture land and by effluent run-off from the storage of animal manure and silage; air pollution from burning crop residues; and soil pollution from agrochemicals (herbicides and pesticides). Insufficient of the energy inputs are utilized and recycled within farm systems, so they become unwanted and damaging outputs, creating an external cost in the environment. The term *external cost* implies the consequences of the damage caused to the environment by the farm sector are met by society at large, and a wide interdisciplinary literature has been developed on the topic (e.g. Brouwer *et al.*, 1991).

Turning to *the modification of 'natural' components within a farming system*, examples include the removal of hedgerows to enlarge field sizes, the drainage of wetlands and the felling of woodlands to create agriculturally productive farmland, and the ploughing of moorland and herb-rich permanent grasslands to create sown pastures of rye grass and clover with higher livestock-carrying capacities (Potter, 1986). In each case the objective is to increase the output of farm products from the farming system. But the external costs include the exposure of large fields to the effects of soil erosion by the wind, the destruction of natural and semi-natural habitats and the ecosystems they support, and the loss of visual amenity in the countryside (i.e. landscape quality). As surveys of the loss of habitats and the incidence of pollution have been completed, so the scale of the environmental damage caused by modern farming systems has been revealed. In the United Kingdom, research by the (former) Nature Conservancy Council revealed losses of 28% of upland woodlands, 51% of lowland marshes, 79% of chalk downlands and 82% of lowland (herb-rich) meadows (Nature Conservancy Council, 1990).

Intervention by the state and agricultural productivism

State intervention in agriculture has increasingly determined the economic context for agricultural production and a useful analytical structure is provided by the following linked sequence: agricultural policy goals, policy instruments and policy impacts (Bowler, 1979).

On *agricultural policy goals*, a distinction can be drawn between utility and equity goals: utility goals concern the contribution by the farm sector towards the performance of the economy; equity goals focus on the provision of satisfactory incomes for the farm population. Column C in Table 4.4 shows the problematic relationship between these two types of goals within the EU. Despite decades of intervention by the state, in most countries the percentage of employment in agriculture still exceeds the percentage of gross national product (GNP) generated by the farm sector, especially in Italy, Spain and Portugal.

Given the complexity of policy goals, it is not surprising that an equally varied set of *agricultural policy instruments* (or measures) has evolved over time. These can be classified into four main groups:

- Instruments to increase demand, e.g. intervention buying, export subsidies, consumer food price subsidies.
- Instruments to reduce supply, e.g. production quotas, land set-aside, farmer retirement pensions.
- Instruments to raise farm incomes by reducing production costs, e.g. fertilizer subsidies, capital investment grants, or by supplementing incomes, e.g. deficiency payments, livestock headage payments.
- Instruments to conserve the natural environment, e.g. woodland planting grants, financial compensation to reduce fertilizer application.

The agricultural policies of most countries contain examples of these mechanisms. Looking at the CAP during the productivist era, intervention agencies were responsible for purchasing surplus farm production at an intervention price, storing that surplus, then exporting the produce through the world market using export subsidies. These mechanisms created the infamous stockpiles of farm produce (wine lakes and butter mountains), the distortion of world trade by export subsidies, and the raising of the cost of intervention to unacceptable levels for taxpayers and consumers in the EU (Bowler, 1985b). Production quotas can be exemplified by the 1984 milk quotas imposed on dairy farmers in an attempt to limit the oversupply of milk. Farmer retirement grants were made available in 1975; headage payments on hill cattle and sheep were provided in the Less Favoured Areas of the EU from 1975 as hill livestock compensatory allowances (HLCAs).

Studies on the *policy impacts* of state intervention have tended to emphasize three features:

- Most policy mechanisms influence the general economic environment for agriculture within which farmers make decisions about what to produce and in what quantity. Thus the state can increase or decrease the element of risk in farmer decision making, thereby 'signalling' the preferred economic behaviour.
- The state can encourage farmers to change their production and farming practices in preferred directions by offering specific financial inducements. But since most inducements are based on voluntary take-up by farmers, the outcome is selective for different groups of farm businesses.
- The state can legislate on specific regulations that dictate farmer behaviour, such as production quotas and set-aside.

Case studies on the impact of productivist policy instruments by agricultural geographers have focused on farm inputs, farm outputs and farm organization. On *farm inputs*, both land and labour inputs have been subjected to intervention by the state. The set-aside of arable land in the European Union is one good example (see below). On the labour input, farmer retirement schemes have been available in many countries for the past three decades, e.g. under the CAP since 1972 (Directive 72/160). The aim has been to increase the rate of reduction in the number of farmers by providing early pensions or lump sum payments. To date, such schemes have been voluntary, with a variable impact between regions. Prosperous farming regions, such as parts of northern France, have been little affected, whereas farmers in regions with low farm incomes have taken advantage of the financial incentives to retire early.

Looking at *farm outputs*, Bowler (1985b) has emphasized how price subsidies from the EU encouraged farmers to specialize even more in those products for which they had an economic advantage. For example, in the productivist era, specialization in wine production increased in the southern regions of France (Languedoc and Rousillon), farmers in south-west Ireland became more specialized in milk production, and farmers in the Paris Basin specialized even further in cereals. Production quotas, in effect a licence to farm a particular product, have had a similar risk-minimizing effect on agriculture. Quotas 'guarantee' an income to the holder, since both the price and quantity of production are known in advance, sometimes for two or three years. Where quotas can be bought and sold, more successful producers are able to purchase entitlements from the less successful; in this way production becomes focused onto fewer farms and regions. The EU milk quota scheme within the United Kingdom has enabled quotas to be sold by farmers in eastern regions to farmers in western regions: thus an historic trend in the regional localization of dairy farming has been deepened.

Studies of policy instruments that affect *the organisation of agriculture* have included farm size and the development of cooperative farming. On farm size, where state intervention in the land market has been permitted – France has a regional network of intervention agencies (SAFER: Sociétés d'Aménagement Foncier et d'Etablissement Rural) – the transfer of land to new owners has been financially expensive and has conferred only marginal benefits on the emerging regional size structure of farms. Indeed, such intervention has been possible on a significant scale only in western and southern regions of France, where the exodus of farm families has been marked and the land cheapest. As for cooperation among farmers, the purchase of farm inputs such as fertilizer and seed, the production of a specific crop or livestock, and the marketing of farm produce, all are encouraged in most countries. Nevertheless, countries vary greatly in the development of farm cooperatives; within western Europe they are strongest in Denmark and the Netherlands.

THEORIZING THE POST-PRODUCTIVIST TRANSITION

Although the exact nature of post-productivist agriculture has yet to be defined by governments and society in developed market economies, the post-productivist transition (PPT) already has a number of known characteristics. These relate especially to EU and US agriculture and include a reduced output of food, the progressive withdrawal of state subsidies, the production of food within an increasingly competitive international market, and the growing environmental regulation of agriculture. The PPT can be interpreted in terms of a progressive reversal of the trends that dominated the preceding productivist era in EU and US agriculture. In particular, it can be conceptualized as three bipolar dimensions of change:

1. *From intensification to extensification* Many policy measures introduced under the CAP since the mid-1980s have encouraged once intensive farm businesses to decrease their level of purchased non-farm inputs and become increasingly more extensive in their production. There are important implications for the reduction in levels of environmental pollution and the restoration of natural habitats, especially in the grass-based livestock and cereal sectors.
2. *From concentration to dispersion* The trend towards increasing polarization in agriculture, whereby most output becomes confined to fewer and larger farm businesses and regions, may be reversed by the 1992 revisions to the CAP (Robinson and Ilbery, 1993). Farmers may be encouraged to subdivide their farm businesses into smaller units, thus dispersing agricultural production. However, there is little current evidence of this, and dispersion is the least likely dimension of change to occur.

3. *From specialization to diversification* Encouraged by the 'cost–price squeeze' on traditional farm products, together with revised policy measures under the CAP, farm businesses are seeking to develop new sources of income through different types of agricultural and non-agricultural diversification. This has necessitated a move away from farming systems where a large proportion of total output is accounted for by a particular product. Such diversification trends enable more diversified land-use systems to be created, with associated implications for the landscape.

The PPT is being strongly regulated by the state; in particular, by three political developments: reforms of the CAP since 1992; the GATT negotiations which were finalized in 1993; and the increasing convergence between agricultural and environmental policy in the EU. One of the main driving forces behind the PPT is global pressure to reduce public policy expenditure and, in the EU, to reduce the percentage of the EU budget devoted to agriculture. But just as the productivist era resulted in a marked differentiation between countries, regions and individual farms, so the PPT can be expected to lead to uneven development and increasing differentiation within the agricultural sector and rural areas (Marsden *et al.*, 1987). Indeed, with the demise of agricultural productivism, new forms of regulation are emerging; they are more localized and detached from national processes associated with productivism, so the emergence of new land uses is inevitable in the PPT (Lowe *et al.*, 1993). Which land uses emerge in particular geographical contexts will partly depend on the different *pathways of farm business development* that are followed by farm households as they adjust to the PPT (Bowler, 1992b).

Six different pathways can be identifed (Table 4.5): pathways 1 and 2 (continuation of profitable food production, possibly aided by biotechnology); pathways 3 and 4 (diversification of the income base through pluriactivity); and pathways 5 and 6 (extensification and semi-retired farming). However, farm households may opt to combine more than one pathway, and three types of farm household can be identified: *accumulators*, *disengagers* and *survivors* (Marsden *et al.*, 1989). Although the accumulators are risk taking and able to redeploy their labour into on-farm

Table 4.5 Pathways of farm business development

---

1. Extension of the industrial model of farm business development based on traditional products and services on the farm

2. Redeployment of farm resources (including human capital) into new agricultural products or services on the farm (agricultural diversification)

3. Redeployment of farm resources (including human capital) into new non-agricultural products or services on the farm (structural diversification)

4. Redeployment of human capital into an off-farm occupation (OGA)

5. Maintenance of traditional farm production and services with reduced capital inputs (extensification)

6. Hobby or part-time (semi-retired) farming

Sources: Bowler (1992b) and Ilbery and Bowler (1993a).

diversification and agri-environmental initiatives, the disengagers lack capital, so they tend to redeploy their labour into other gainful activities (OGAs) off the farm. The survivors attempt to continue with traditional farming and may become passive adopters of certain environmental schemes (Morris and Potter, 1995); any OGAs developed will tend to be agriculturally related, such as agricultural contracting.

EMPIRICAL EVIDENCE OF THE POST-PRODUCTIVIST TRANSITION

Surplus farmland and set-aside

Set-aside is an example of extensification (pathway 5). Land retirement programmes have long been in existence in the United States, where they have met with limited success (Potter *et al.*, 1991). Major problems with set-aside in America have included *slippage* and *selectivity* effects. Slippage occurs when the fall in production is proportionately lower than the amount of land set aside, causing the overall impact to be less than expected. Slippage of up to 50% can be caused by farmers retiring their least-productive land first and intensifying production on the remaining land. The selectivity effect involves farmers participating in schemes to subsidize farm changes that they would have made anyway. This is especially the case with farmer retirement, but also applies to farmers interested in farm woodland and conservation, and other restructuring strategies (Jones *et al.*, 1993).

Despite knowledge of such problems, the EU introduced the set-aside of arable land in 1988 as part of a package of measures designed to cut overproduction in member states. The initial five-year scheme was compulsory for all member states, with the exception of Portugal, but voluntary for individual farmers. In return for setting aside at least 20% of their arable land for a minimum of five years, farmers received compensation of between 100 and 600 ecu per hectare per year. Retired land had to be left fallow (with rotational and some grazing possibilities), wooded, or used for non-agricultural purposes such as recreation and conservation. An additional one-year set-aside scheme was introduced in 1991.

The initial response to set-aside in the EU was disappointing (Ilbery, 1992). After two years, just 798 508 hectares were retired, representing 1.3% of the arable land and 2% of the area devoted to cereals in the EU. Even by 1992 less than 2 million hectares (2.6% of arable land) had been retired (Table 4.6), compared with conservative estimates of the need to retire at least 6 million hectares. Nationally, the uptake of set-aside was very uneven, with Italy and the former West Germany accounting for over 60% of the total. Such differences reflected national attitudes towards set-aside and variable rates of compensation. For example, although West Germany and the United Kingdom were early supporters of set-aside, Denmark and France were opposed to the idea. Rates of compensation varied accordingly and, although some countries offered flat rates of compensation, like Ireland, Luxembourg (both low) and the Netherlands (high), others offered variable premiums based on different criteria and areas (e.g. land quality, irrigated areas, Less Favoured Areas).

Table 4.6 Distribution of set-aside in the European Union

Country	1988–1992[a]		1993/94[b]	
	Area (ha)	Percent of EU total	Area (000 ha)	Percent of EU total
Belgium	880	0.05	19	0.41
Denmark	12 813	0.74	208	4.52
Germany	479 260	27.78	1 050	22.80
Greece	713	0.04	15	0.33
Spain	103 169	5.98	875	19.00
France	235 492	13.65	1 578	34.27
Italy	721 847	41.82	195	4.23
Ireland	3 452	0.20	26	0.56
Luxembourg	91	0.01	2	0.04
Netherlands	15 373	0.89	8	0.17
Portugal	exempt		61	1.32
United Kingdom	152 700	8.85	568	12.33
Total	1 725 790	100.00	4 605	100.00

[a] Voluntary set-aside.
[b] Compulsory set-aside.
Source: DG VI, European Commission.

Rates of compensation suggested that set-aside would be more attractive in the marginal rather than main cereal-growing regions. Briggs and Kerrell (1992) confirmed this at the EU level, as did Jones (1991) and Ilbery and Bowler (1993a) at the regional level in West Germany and the United Kingdom, respectively. In England, for example, set-aside was not concentrated in the 'core' cereal-producing areas of East Anglia and Humberside, but in those (marginal) areas with a weak competitive ability in cereals and a strong competitive ability in both mixed farming and part-time farming (Bowler and Ilbery, 1989). This meant a relative concentration of set-aside in the South-East and in close proximity to London (Fig. 4.2a), possibly reflecting market opportunities for OGAs and diversification activities (Ilbery, 1990). Indeed, Jones *et al.*, (1993) confirmed that, as well as being concentrated in marginal cereal areas in Rheinland-Pfalz, German farmers were using set-aside to restructure their farm businesses. In particular, it was helping to secure generational continuity of the farm business, to postpone new rounds of investment, to secure non-agricultural income, and to act as a pre-retirement strategy to run down the holding (pathway 6).

With cereal output not falling in the EU, the twin problems of slippage and selectivity had not been resolved. Consequently, set-aside became compulsory for all commercial cereal farmers as part of the CAP reforms in 1992. In return for setting aside arable land, farmers became eligible to claim income aid through an Arable Areas Payments Scheme (AAPs). Two main set-aside options are now available to farmers: rotational and non-rotational. Rotational set-aside requires

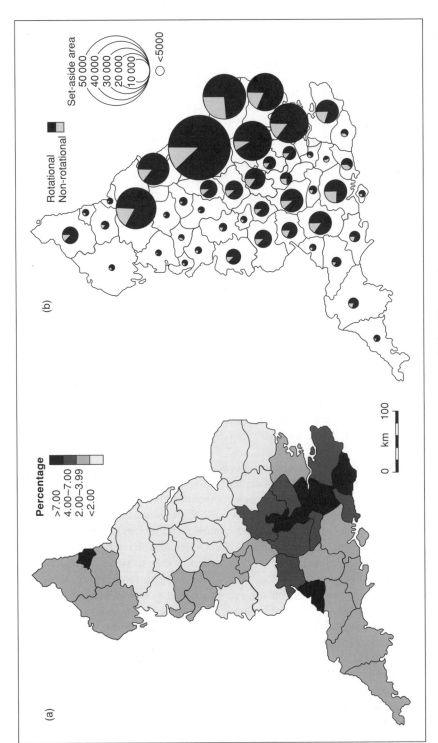

Figure 4.2 Distribution of set-aside in England: (a) 1991 voluntary scheme and (b) 1993 compulsory scheme

farmers to set aside 15% of their arable land on a rotational basis for each of six years, whereas non-rotational set-aside involves setting aside the same 18% of arable land for five years.

In 1993, 4.6 million hectares were set aside in the EU (Table 4.6); by 1994 this had risen to 5.9 million hectares. In England, less than 20% of the 553 175 hectares set aside in 1994 was non-rotational, but at least the 'core' cereal areas are now setting aside more arable land (Fig. 4.2b). Indeed, cereal output in the EU has been falling since 1992 (except 1996) and there has been a concomitant fall in the purchase of fertilizers and pesticides. Nevertheless, set-aside continues to be criticized for its lack of environmental concern (Craighill and Goldsmith, 1994). EU countries are encouraged to consider long-term set-aside as part of their agri-environmental package, and the United Kingdom has introduced a voluntary 20-year Habitat Improvement Scheme to present an opportunity for constructive use of set-aside land (Chapter 5).

Pluriactivity

Pluriactivity refers to the generation, by farm household members, of income from on-farm and/or off-farm sources in addition to income obtained from primary agriculture. Thus it involves both *farm diversification* and *other gainful activities* (OGAs), or pathways 3 and 4 in Table 4.5. Indeed, the two elements of pluriactivity were originally examined in isolation, with farm diversification being classified into two types (Ilbery, 1991): agricultural (e.g. unconventional crop and livestock enterprises, woodland and organic farming) and structural (e.g. farm-based tourism, adding value through direct marketing and/or processing, and craft/light industries). However, since the late 1980s, research has been directed towards the wider concept of pluriactivity (Fuller, 1990; MacKinnon et al., 1991).

The incidence of pluriactivity in many developed market economies is high and increasing during the PPT. In France the proportion of farm household income coming from pluriactivity increased from 15% in 1956 to 42% in 1988 (Benjamin, 1994). A major research project conducted by the Arkleton Trust in 24 regions of western Europe in the late 1980s indicated that 58% of farm households were pluri-active (Fuller, 1990). The greatest proportion of pluriactivity was off-farm (55% of all households), but this varied significantly between regions, from a high of 81% in Freyung-Grafenau (Germany) and 72% in West Bothnia (Sweden) to 27% in Picardie (France) and 33% in Andalucia (Spain). More recent research has confirmed the high proportion of pluriactive households; in all instances, OGAs have been more numerous than on-farm diversification and they tend to be dominated by farm wives (Le Heron et al., 1994). Not surprisingly, therefore, the Arkleton Trust project indicated that one-third of farm households obtained more than 50% of their income from off-farm sources in the late 1980s. However, this varied from just 10% in Picardie to 71% in Freyung-Grafenau.

Farm tourism, especially farm-based accommodation, is often the dominant type of on-farm diversification (Evans and Ilbery, 1992). The incidence of alternative crops and livestock (including organic farming and woodland) is not very high,

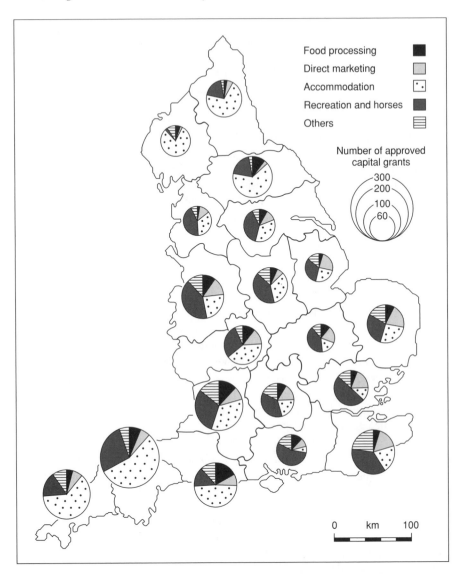

Figure 4.3 Uptake of the Farm Diversification Grant Scheme in England (Reprinted from Ilbery and Stiell, 1991, by permission of the Geographical Association)

although more research is needed on this topic. Different areas present different opportunities and constraints for farm diversification. For example, the uptake of the Farm Diversification Grant Scheme in England was biased towards urban fringe and marginal fringe regions, especially in the South (Ilbery and Stiell, 1991). Urban fringe areas attracted farm-based recreational activities, especially 'horsiculture', whereas marginal (mainly tourist) areas encouraged farm-based accommodation (Fig. 4.3).

However, geographical patterns of pluriactivity are rarely that straightforward. Instead, they reflect the interaction of a number of factors external and internal to the farm business. For example, a relationship exists between regional socio-economic conditions (e.g. local labour markets, unemployment) and rates of pluriactivity. Thus, pluriactivity is further developed in those regions where labour markets are well structured and diverse. Nevertheless, the evidence is not conclusive and other factors have to be considered. Although forces beyond agriculture explain the incidence and patterns of pluriactivity, the growing participation of women in the labour force for social rather than economic reasons has become a dominant driving force. Similarly, local cultural factors and specific landscape designations, such as national parks, can affect the distribution of pluriactivity (Bateman and Ray, 1994).

Different types of pluriactivity are affected by different combinations of factors. Although the distribution of on-farm diversification may reflect regional economic conditions, the distribution of OGAs may not. In a recent survey of Scottish pluriactivity, OGAs proved to be most important in the least densely populated areas, reflecting social and cultural factors instead of economic factors (Edmond and Crabtree, 1994). Indeed, the distribution of OGAs can often be better explained in terms of personal preferences and household characteristics.

Household characteristics, especially stages in the family life cycle, belong to a set of internal factors which further complicate the geography of pluriactivity. Research in England and Wales has indicated that many 'adopters' of farm diversification share certain characteristics which distinguish them from 'non-adopters' (Ilbery and Bowler, 1993b). For example, they tend to have larger farm businesses, higher net incomes and a greater degree of indebtedness. Similarly, adopters tend to be younger than non-adopters and have continued their full-time education after school and received more formal agricultural training. Significantly, a greater proportion of adopters have children wishing to continue the farm business, thus acting as a stimulus to the development of farm diversification. Size of farm is an important factor in pluriactivity. In particular, OGAs dominate farm households with small farms, whereas farm diversification is biased towards larger farms (Gasson, 1988). Similar associations have been found between pluriactivity and farm type, tenure, education, household composition and especially indebtedness.

It is the interaction of internal factors with the socio-economic environment external to the farm household which shapes broad patterns of pluriactivity and particular concentrations of farm diversification and OGAs. This has been demonstrated in a study of pluriactivity in three French agricultural regions by Campagne *et al.* (1990). In Picardie, an area of large farms and specialist arable production in northern France, household members are using agricultural resources to increase non-agricultural activities. Many wives work outside agriculture and there is an enterprise culture among the mainly family-based farms. Campagne *et al.* describe this as a zone of *business pluriactivity*.

Languedoc, by way of contrast, is a wine-growing region in southern France which has a long history of pluriactivity. Income generated from mainly off-farm sources, especially by the spouse and children, is used to modernize and maintain the farm business; this is described as the *pluriactivity of maintaining farming*. Finally, the Savoy valleys in south-east France are physically more marginal for agriculture, and farm households combine farming with a diverse range of income-generating activities on and off the farm. With the progressive abandonment of farming, pluriactivity becomes a survival mechanism, hence the phrase *pluriactivity for survival*.

Pluriactivity, in its many different forms, is an adjustment strategy being adopted by many farm households during the PPT. Several factors, both internal and external to the farm, help to account for its geographically uneven development.

Environmental goods

The third example of the move towards post-productivism is provided by payments to farmers for the purpose of environmental protection and countryside management. A common thread linking the various *agri-environmental schemes* is that they are mainly voluntary, rely on incentive payments, and are limited to a restricted number of designated areas. A detailed account of agri-environmental policy development and change is provided in Chapter 5, so it will not be reproduced here. However, it is worth emphasizing that agri-environmental measures are unlikely to be fully effective while remaining 'bolted-on' to an economically driven policy (Robinson, 1991).

Indeed, Potter (1993) estimated that just 2% of the farm budget is likely to be spent on agri-environmental schemes in the United Kingdom, with the remaining 98% being reserved for supporting prices and funding compensation measures like set-aside. Moreover, outside the designated areas, variations to the productivist model of farming depend on decision making by individual farm households. For example, farmers can vary their decisions on whether to place set-aside land into short-term (rotational) or long-term (conservation) uses, including a range of non-food crops such as biofuels. Similarly, farmers have voluntary entry into other state-financed schemes to promote the provision of 'environmental goods'; UK schemes include Habitat, Moorland, Farm Woodland Premium, Countryside Stewardship and Countryside Access. Patterns of response will vary between individual farmers and farming regions.

THE PROSPECTS FOR A SUSTAINABLE AGRICULTURE

The post-productivist transition has, at its core, the concept of a sustainable agriculture. The origins of this now popularized term can be traced to the vocal and politically influential environmental movement which, from the 1960s and 1970s, turned its attention to the damaging environmental consequences of productivist agriculture. Beginning with the ecological consequences of using pesticides such as DDT (Carson, 1963), concern was extended to the wider external costs of agricultural practices, including in more recent years the declining health standards of food and animal welfare. The term *sustainable* is now being applied to agriculture, as to all aspects of human existence, following the widely cited definition promulgated in the Brundtland Report by the World Commission on Environment and Development (WCED):

> development that meets the needs of the present without compromising the ability of future generations to meet their own needs. (World Commission on Environment and Development, 1987)

Political impetus has been given to the development of sustainable agriculture by the international agreement on Agenda 21 at the 1992 Earth Summit at Rio de Janeiro and its subsequent interpretation in national contexts (e.g. the UK Sustainable Development Report of 1993).

The meaning of sustainable agriculture

A considerable literature has emerged in a relatively short period of time on the problem of defining *sustainable* in a rural context (e.g. Brklacich *et al.*, 1990; Moffatt, 1992). Pierce (1992) has traced the intellectual and academic precursors of the term from writers in the eighteenth century to the present and three basic propositions can be identified:

- Rates of use of renewable resources should not exceed their rates of regeneration.
- Rates of use of non-renewable resources should not exceed the rate at which sustainable renewable substitutes are developed.
- Rates of pollution emission should not exceed the assimilative capacity of the environment.

Within these propositions, a distinction can be drawn between natural capital, manmade capital and human capital, and between strong and weak sustainability. Strong sustainability permits no depletion in any single type of capital (i.e. no substitutability), whereas weak sustainability allows depletion in one type of capital to be substituted by increments in either of the alternative types of capital, thereby maintaining the overall stock of capital (i.e. constant capital). In this way, dynamic change in an agricultural system can be accounted for. According to this argument, weak sustainability can be achieved by depleting natural (i.e. biophysical) capital in line with a balancing increase in manmade or human capital. A premium is placed on the pricing of all elements in a defined agricultural system so as to be able to measure the substitutability effect, including the 'valuing' of the natural environment with a distinction drawn between 'use' and 'existence' values.

Brklacich *et al.* (1990) have attempted to translate such propositions into more tangible agricultural terms; for them sustainable agriculture simultaneously provides three things:

- *Environmental sustainability* is the capacity of an agricultural system to be reproduced into the future without unacceptable pollution, depletion or physical destruction of its natural resources such as soil, water, air and natural or semi-natural habitats.

■ *Socio-economic sustainability* is the capacity of an agricultural system to provide an acceptable economic return to those employed in the productive system.
■ *Productive sustainability* is the capacity of an agricultural system to supply sufficient food to support the non-farm population.

All of these dimensions await the definition of terms such as *acceptable* and *sufficient*, along with concepts such as the sustainable level of state intervention in support of agriculture, and the spatial scope of an agricultural system as regards its place in an international system rather than merely regional or national sustainable food supply systems.

Alternative models of sustainable agriculture

The problem facing post-productivist agriculture is the translation of broad concepts on sustainability into tangible agricultural systems. Moffatt (1992) puts it succinctly: 'The task ahead is to translate . . . noble sentiments into deeds'. But this very translation is contested (Bowler and Ilbery, 1993): from an idealist (ecocentric or alternative agriculture) viewpoint the organization of socio-economic systems into 'no' or 'low' growth modes is the only viable long-term option for human society, with implications for radical changes in consumption patterns, resource allocation and utilization, and individual lifestyles. From this viewpoint a checklist of desirable attributes of a sustainable agriculture includes diversified land use, integration of crop and livestock farming, crop rotations, the use of organic manure, nutrient recycling, low energy inputs as well as low inputs of agrochemicals, and biological disease control. The resulting agricultural models of sustainable systems include organic agriculture, permaculture, low input–output farming, alternative agriculture, regenerative farming, biodynamic farming and ecological farming.

An instrumentalist (technocentric or conventional agriculture) viewpoint contests the utility of these goals as being practically and politically unrealistic; instead *sustainable* is interpreted as a contextual process, rather than a set of prescriptions for exactly how an agricultural system should operate, recognizing that ideal models of sustainable agriculture are unlikely ever to be achieved. Instrumentalist agricultural models include integrated crop management (ICM), diversified agriculture, extensified agriculture and conservation agriculture. The danger of the instrumentalist viewpoint is that sustainability can become a politically malleable aspiration rather than a rigorous programme of target setting and policy implementation.

To date in the PPT, a minority of farmers are pursuing idealist models of sustainable agricultural development, as discussed earlier for organic farming. Although these models provide advantages as regards the elimination of agrochemicals and inorganic fertilizers, and create an increased demand for farm labour, they have come under criticism for their low productivity in terms of food yield per hectare of farmland, and for pollution of watercourses through run-off from organic nutrients. Rather, the majority of farmers, under increased regulation by the state, are following instrumentalist models of sustainable agricultural development, with limits being placed on the application of fertilizers, constraints on the types and rates of application of agrochemicals, the imposition of higher minimum standards of pesticide residues in food products, and subsidies to farm under lower input–output systems.

CONCLUSION

Three main points can be made by way of a brief conclusion. First, the differentiation of agriculture under productivism is continuing under the PPT, with new divisions in agriculture developing as regions become dominated by different pathway choices. Secondly, productivist agriculture is likely to be more persistent than many critical researchers appear to assume, and the deepening dualism between productivist and PPT farms is set to continue. Thirdly, PPT elements on capitalist farms are being brought about by state regulation, whereas a greater participation and collaboration between capitalist farms and local groups and institutions may be a more effective mechanism for the mid-term.

REFERENCES

Bateman, D. and Ray, C. (1994) Farm pluriactivity and rural policy: some evidence from Wales. *Journal of Rural Studies*, **10**, 1–13.

Benjamin, C. (1994) The growing importance of diversification activities for French farm households. *Journal of Rural Studies*, **10**, 331–42.

Bowler, I.R. (1979) *Government and agriculture: a spatial perspective.* Longman, London.

Bowler, I.R. (1985a) Some consequences of the industrialisation of agriculture in the European Community. In Healey, M.J. and Ilbery, B.W. (eds) *The industrialisation of the countryside.* GeoBooks, Norwich, pp. 75–98.

Bowler, I.R. (1985b) *Agriculture under the Common Agricultural Policy.* Manchester University Press, Manchester.

Bowler, I.R. (ed) (1992a) *Geography of agriculture in developed market economies.* Longman, London.

Bowler, I.R. (1992b) Sustainable agriculture as an alternative path of farm business development. In Bowler, I.R., Bryant, C.R. and Nellis, M.D. (eds) *Rural systems in transition: agriculture and environment.* CAB International, Wallingford, pp. 237–53.

Bowler, I.R. and Ilbery, B.W. (1989) The spatial restructuring of agriculture in the English counties. *Tijdschrift voor Economische en Sociale Geographie*, **80**, 302–11.

Bowler, I.R. and Ilbery, B.W. (1993) Sustainable agriculture in the food supply system. In Nellis, M.D. (ed) *Geographical perspectives on the social and economic restructuring of rural areas.* Kansas State University, Manhattan KS, pp. 4–13.

Briggs, D.J. and Kerrell, E. (1992) Patterns and implications of policy-induced agricultural adjustments in the European Community. In Gilg. A.W. (ed) *Restructuring the countryside: environmental policy in practice.* Avebury, Aldershot.

Brklacich, M., Bryant, C. and Smit, B. (1990) Review and appraisal of concepts of sustainable food production systems. *Environmental Management*, **15**, 1–14.

Brouwer, F., Thomas, A. and Chadwick, M. (eds) (1991) *Land use changes in Europe.* Kluwer Academic, Dordrecht, pp. 49–78.

Campagne, P., Carrère, G. and Valceschini, E. (1990) Three agricultural regions of France: three types of pluriactivity. *Journal of Rural Studies*, **4**, 415–22.

Carson, R. (1963) *Silent spring*. Hamish Hamilton, London.

Commission of the European Communities (various years) *The agricultural situation in the Community*. Annual Report, EC, Brussels.

Craighill, A. and Goldsmith, E. (1994) A future for set-aside? *Ecos*, **15**, 58–62.

Edmond, H. and Crabtree, R. (1994) Regional variation in Scottish pluriactivity: the socio-economic context for different types of non-farming activity. *Scottish Geographical Magazine*, **110**, 76–84.

Evans, N.J. and Ilbery, B.W. (1992) Farm-based accommodation and the restructuring of agriculture: evidence from three English counties. *Journal of Rural Studies*, **8**, 85–96.

FitzSimmons, M. (1986) The new industrial agriculture: the regional differentiation of speciality crop production. *Economic Geography*, **62**, 334–53.

Friedmann, H. (1986) Family enterprise in agriculture: structural limits and political possibilities. In Cox, G., Lowe, P. and Winter, M. (eds) *Agriculture: people and politics*. Allen and Unwin, London, pp. 20–40.

Fuller, A.J. (1990) From part-time farming to pluriactivity: a decade of change in rural Europe. *Journal of Rural Studies*, **6**, 361–73.

Gasson, R. (1988) Farm diversification and rural development. *Journal of Agricultural Economics*, **39**, 175–82.

Goodman, D.E. and Redclift, M.R. (eds) (1989) *The international farm crisis*. Macmillan, London.

Goodman, D.E. and Redclift, M.R. (1991) *Refashioning nature: food, ecology and culture*. Routledge, London.

Goodman, D.E., Sorj, B. and Wilkinson, J. (1987) *From farming to biotechnology: a theory of agro-industrial development*. Blackwell, Oxford.

Gregor, H.F. (1982) *Industrialisation of US agriculture: an interpretative atlas*. Westview CO.

Hart, P.W. (1992) Marketing agricultural produce. In Bowler, I.R. (ed) *Geography of agriculture in developed market economies*. Longman, London, pp. 162–206.

Ilbery, B.W. (1990) Adoption of the arable set-aside scheme in England. *Geography*, **76**, 259–63.

Ilbery, B.W. (1991) Farm diversification as an adjustment strategy on the urban fringe of the West Midlands. *Journal of Rural Studies*, **7**, 207–18.

Ilbery, B.W. (1992) Agricultural policy and land diversion in the European Community. In Gilg, A.W. (ed) *Progress in Rural Policy and Planning*, vol 2. Belhaven, London, pp. 153–66.

Ilbery, B.W. and Bowler, I.R. (1993a) Land diversion and farm business diversification in EC agriculture. *Nederlandse Geografische Studies*, **172**, 15–27.

Ilbery, B.W. and Bowler, I.R. (1993b) The Farm Diversification Grant Scheme: adoption and non-adoption in England and Wales. *Environment and Planning C*, **11**, 161–70.

Ilbery, B.W. and Stiell, B. (1991) Uptake of the Farm Diversification Grant Scheme in England. *Geography*, **75**, 69–73.

Jansen, A. and Hetsen, H. (1991) Agricultural development and spatial organisation in Europe. *Journal of Rural Studies*, **7**, 143–51.

Jones, A. (1991) The impact of EC set-aside policy: the response of farm businesses in Rendsburg-Eckernforde (Germany). *Land Use Policy*, **8**, 108–25.

Jones, A., Fasterding, F. and Plankl, R. (1993) Farm household adjustments to the European Community's set-aside policy: evidence from Rheinland-Pfalz (Germany). *Journal of Rural Studies*, **9**, 65–80.

Le Heron, R. (1993) *Globalised agriculture: political choice*. Pergamon, Oxford.

Le Heron, R. Roche, M. and Johnston, T. (1994) Pluriactivity: an exploration of issues with reference to New Zealand's livestock and fruit agro-commodity systems. *Geoforum*, **25**, 155–71.

Lowe, P., Murdoch, J., Marsden, T., Munton, R. and Flynn, A. (1993) Regulating the new rural spaces: the uneven development of land. *Journal of Rural Studies*, **9**, 205–22.

MacKinnon, N., Bryden, J., Bell, C., Fuller, A. and Spearman, M. (1991) Pluriactivity, structural change and farm household vulnerability in western Europe. *Sociologia Ruralis*, **31**, 58–71.

Marsden, T. and Symes, D. (1984) Land ownership and farm organisation: evolution and change in capitalist agriculture. *International Journal of Urban and Regional Research*, **8**, 388–401.

Marsden, T., Munton, R., Whatmore, S. and Little, J. (1986) Towards a political economy of capitalist agriculture: a British perspective. *International Journal of Urban and Regional Research*, **10**, 498–521.

Marsden, T., Whatmore, S. and Munton, R. (1987) Uneven development and the restructuring process in British agriculture: a preliminary exploration. *Journal of Rural Studies*, **3**, 297–308.

Marsden, T., Munton, R., Whatmore, S. and Little, J. (1989) Strategies for coping in capitalist agriculture: an examination of responses of farm businesses in British agriculture. *Geoforum*, **20**, 1–14.

Moffatt, I. (1992) The evolution of the sustainable development concept: a perspective from Australia. *Australian Geographical Studies*, **30**, 27–42.

Morris, C. and Potter, C. (1995) Recruiting the new conservationists. *Journal of Rural Studies*, **11**, 51–63.

Nature Conservancy Council (1990) Nature conservation and agricultural change, *Focus on Nature Conservation*, **25**. Nature Conservancy Council, Peterborough.

Newby, H. (1982) Rural sociology and its relevance to the agricultural economist: a review. *Journal of Agricultural Economics*, **33**, 125–65.

Pierce, J.T. (1992) Progress and the biosphere: the dialectics of sustainable development. *Canadian Geographer*, **36**, 306–20.

Potter, C. (1986) Processes of countryside change in lowland England. *Journal of Rural Studies*, **2**, 187–95.

Potter, C. (1993) Pieces in a jigsaw: a critique of the new agri-environment measures. *Ecos*, **14**, 52–54.

Potter, C., Burnham, P., Edwards, A., Gasson, R. and Green, B. (1991) *The diversion of land: conservation in a period of farming contraction*. Routledge, London.

Redclift, M. (1984) *Development and the environmental crisis: red or green alternatives?* Methuen, London.

Robinson, G.M. (1991) EC agricultural policy and the environment: land use implications in the UK. *Land Use Policy*, **8**, 301–11.

Robinson, G.M. and Ilbery, B.W. (1993) Reforming the CAP: beyond MacSharry. In Gilg, A.W. (ed) *Progress in Rural Policy and Planning*, vol 3. Belhaven Press, London, pp. 197–207.

Shucksmith, M. (1993) Farm household behaviour and the transition to post-productivism. *Journal of Agricultural Economics*, **44**, 466–78.

Traill, B. (1989) *Prospects for the European food system*. Elsevier, London.

Troughton, M.J. (1986) Farming systems in the modern world. In Pacione, M. (ed), *Progress in agricultural geography*. Croom Helm, London, 93–123.

Vandergeest, P. (1988) Commercialization and commoditization: a dialogue between perspectives. *Sociologia Ruralis*, **28**, 7–12.

Whatmore, S. (1995) From farming to agribusiness: the global agro-food system. In Johnston, R.J., Taylor, P.J. and Watts, M.J., *Geographies of global change: remapping the world in the late twentieth century*. Blackwell, London, 3–49.

Wrigley, N. (1987) The concentration of capital in UK grocery retailing. *Environment and Planning A*, **19**, 1283–88.

World Commission on Environment and Development (1987) *Our common future*. Oxford University Press, Oxford.

CHAPTER 5

CONSERVING NATURE:
agri-environmental policy development and change
Clive Potter

INTRODUCTION

The arguments of conservationists have been part of the rural policy debate for some time now. With its roots in Romantic protest and transcendentalism, ideas of conservation long predate 1960s environmentalism, crystallizing in the early years of the century into a system that describes how best to manage water and soil resources and protect landscapes, habitats and species. The first agencies designed to promote soil conservation were set up in the United States during the Great Depression of the 1930s and there has been an unbroken history of government intervention in this field ever since.

In a country like Britain, nature and landscape protection as activities of government date from the late 1940s, when in a welter of postwar reconstructionist legislation, policies for the selection, protection and management of national parks and a series of national nature reserves were established along with the first scientifically grounded environmental agency in the world, the Nature Conservancy.

By common consent, however, it is only since the mid-1980s, when steps were taken to implement a wave of 'agri-environmental' policy schemes and programmes, that conservation has become a force for rural land-use change in either country. Why is this? Part of the explanation lies in the way now outdated assumptions about agricultural change and the relationship between agriculture and the environment were written into traditional conservation strategies and programmes.

British nature conservationists could afford the luxury of focusing their efforts on what Adams (1986) calls 'conservation stamp collecting' because nature was not regarded as threatened elsewhere. Indeed, agricultural expansion was positively the guarantor of a diverse and beautiful countryside. In the United States, this doctrine of agri-environmental compatibility was just as influential in sanctioning soil conservation programmes that have arguably been more successful in supporting the incomes of farmers than conserving the soil.

By the late 1980s the picture was undergoing radical change. The modernization and intensification of farming that have been such a defining feature of the past 30 years (Chapter 4) shattered many of these core beliefs and encouraged conservationists to challenge the way farm policies operated. With the first moves towards environmental reform of farm policies in both the United States and the European Union (EU), conservation concerns were beginning to influence the management of rural land on more farms and in many more locations than ever before, and conservation thinking was becoming more ambitious and expansive in scope. No longer an attitude of defence, nature conservation in the United Kingdom was rebalanced to extend protection to entire agricultural landscapes and there were the beginnings of a debate about recreating and restoring some of what had been lost in the recent past. American concerns also widened to take in questions of water quality and wetland protection, swinging away from the pursuit of soil conservation for its own sake in favour of a broader agenda which involved environmentalists taking their concerns into the agricultural policy field.

In reviewing what has been achieved and speculating about the future, this chapter sets out to assess the broader significance of these important developments in conservation thought and action. It begins by rehearsing the traditional policy stance taken towards resource and nature conservation in the United Kingdom and the United States, and its inadequancies in the face of an unprecedented intensfication of agricultural production during the postwar period. The chapter goes on to describe the important shift in public attitudes towards farming which occurred during the late 1970s and early 1980s, and the impetus this gave to agricultural policy reform. It goes on to analyse the resulting agri-environmental policy initiatives and their increasingly significant influence over rural land-use change. The chapter concludes with a discussion of the new strategies which conservationists are now set to pursue in a period of international farm policy retrenchment and the 'new conservation' which could be the result.

LAYING THE FOUNDATIONS

It is tempting to view the first steps towards resource conservation – in the United States during the 1930s and a decade later in Britain – as an inevitable response to growing public concern. In fact, as both Sheail (1994, 1995) and Hays (1959, 1987) show for their respective countries, it was chiefly due to the exertions of a few key individuals that conservation came to be seen as a government responsibility at all. In the United States, the soil scientist Hugh Hammond Bennett succeeded in gaining the attention of Congress and the Roosevelt administration by dramatizing the soil erosion threat to agricultural productivity and farming livelihoods. As Hays (1959) points out, Bennett's authority as a scientist had immense appeal to a president keen to pursue government by experts. But more than this, state-financed soil conservation was politically appealing because it offered a means of supporting farmers hard hit by the Depression that was by then threatening communities throughout rural America, in the Midwest especially.

The Great Depression, in fact, created the economic and social context within which government action on a large scale became possible and desirable, allowing policy makers finally to put into practice the interventionist programmes for resource conservation formulated earlier in the century by progressives like Theodore

Roosevelt and Gifford Pinchot. They began with the creation of the Soil Conservation Service (SCS) in 1935, charged with providing advice and assistance to landowners. But the more significant policy development came in 1936 when Congress passed an Act allowing the United States Department of Agriculture (USDA) to pay farmers to take land growing surplus 'soil depleting' crops out of production. The perception that soil conservation and the support of farmers' incomes could be pursued through the same set of policies was a critical one and reflected a long-held progressive conservationist belief that farmers were managers of nature, extracting a bounty to ensure the continued prosperity of the nation (Batie, 1988).

Because soil erosion was regarded as a symptom of depressed farm incomes, part of the solution, it was assumed, lay in restricting the overproduction which was its root cause. Conservation practices such as terracing, intercropping and developing grass waterways were just as effective in reducing the supply of agricultural products to an already saturated market as conserving soil. They also allowed USDA to disburse many millions of dollars to farmers under the guise of the Agricultural Conservation Programme (ACP) which, as Reichelderfer (1992, p. 5) observes, 'was designed with agricultural interests in mind . . . [it] was easily integrated with concurrent efforts to adjust agricultural production and stabilise prices'. Since the 1930s it has been estimated that USDA has invested over $15 billion on 'soil conserving practices' on the nation's farms. By 1980 some 2.4 million farmers – practically the entire US farm population – belonged to soil conservation districts and 364 000 were actively participating in the ACP (Brubaker and Castle, 1982).

According to Max Nicholson, quoted in Sheail (1995), it was only with the suspension of the 'normal mechanisms of ensuring inaction' under conditions of war that the arguments of British nature conservationists like Sir Julian Huxley and A.G. Tansley began to get a hearing in Whitehall. In the depths of war, nature conservation and the movement for national parks became one of a series of ideas deployed to raise morale and focus attention on reconstructon. A blueprint for nature conservation had already been laid down by a committee appointed by the Minister for Town and Country Planning under the chairmanship of Huxley himself. The priority was to select and designate a series of nature reserves and other protected sites which, in the words of the committee, would contain 'a balanced representation of the different major types of plant and animal communities in England and Wales, while at the same time including certain unique sites of the highest value to science' (Huxley, 1947, p. 7).

Yet, as Felton (1993) has pointed out, the luxury of being able to present nature conservation as a largely scientific and educational project centred on the selection, management and protection of special sites, was only possible because a prescribed form of countryside management was perceived not to be required elsewhere. This was the message of the influential Scott Committee of 1942 on land utilization in rural areas, which in a much quoted phrase had argued that 'farmers are unconsciously the nation's landscape gardeners'. The Scott philosophy became the touchstone of postwar planning, justifying a dual approach to rural land-use management which combined strict control over urban development to safeguard productive agricultural land with an expansionist farm support policy bereft of any environmental safeguards.

For the next 30 years British nature conservationists would devote much of their resources and expertise to creating a conservation estate made up of national nature reserves and sites of special scientific interest; with their few relevant powers, they would show little inclination to influence the course of change in the wider countryside. The prevailing image was of a pyramidal countryside, with these key sites at its apex, resting on a broad base of 'good wildlife habitat' assumed to exist elsewhere. Felton (1993) comments how this analogy, repeatedly used in official publications, merely reinforced the impression that agricultural landscapes, if not expendable, were certainly of lesser value in conservation terms. They could safely be left to the farmers' care.

AGRI-ENVIRONMENTALISM

By the late 1970s, however, evidence was beginning to accumulate which challenged the environmental credentials of farmers in both the United States and the European Union, forcing a profound reassessment of conservation policies and programmes in these countries. Indeed, the shift in public opinion was to be so pronounced that, instead of enjoying a reputation as stewards of the countryside and its natural resources, farmers found themselves castigated as environmental destroyers responsible for 'the theft of the countryside' (Shoard, 1984). Not that recognition was instantaneous. In the United Kingdom and other western European countries there was a noticeable lag between publication of the first scientific studies documenting various environmental impacts and the development of public interest and concern. Nonetheless, the 1970s saw a growing conviction that many of the adverse environmental effects being observed were intrinsic or at least strongly associated with modern farming techniques and changes in farming systems.

Throughout western Europe the intensification and growing specialization of farming has operated on a number of levels, fragmenting and simplifying many farmed landscapes through the removal of hedgerows, trees, small woods, ponds and the reclamation of wetland and unimproved grassland. In what Meeus *et al.* (1990) describe as the 'former enclosure landscapes' of lowland England, north-west Denmark and in the classic *bocage* of northern France and Belgium, the removal of hedgerows from the late 1960s onwards to facilitate larger-scale farming operations was the most visible sign of gathering agricultural change. Hedgerow loss in Britain was running at over 8000 kilometres per year throughout the 1970s, whereas between 1987 and 1984 there was an estimated cumulative loss of 28 000 kilometres (Barr *et al.*, 1993).

From a nature conservation perspective, this and other symptoms of change meant a substantial reduction in the biodiversity of agricultural land and the increasing ecological isolation of the nature reserves and other key sites which remained. In 1979 the Nature Conservancy Council (NCC) made this bleak assessment of the prospects for wildlife in the wider countryside: 'While a few habitats that are rich in wildlife are increasing, most often in the intensively farmed parts of Britain, they are declining in size, in quality or both. The decline is serious: it is occurring throughout the lowlands and more fertile uplands of England, Wales and Scotland ... the rate and extent of change during the last 35 years have been greater than at any similar length of time in history' (Nature Conservancy Council, 1979, p. 21). A substantial loss of seminatural vegetation in more remote areas, as land was reclaimed and improved to support expanded livestock numbers, was particularly serious from a nature conservation point of view.

It has been estimated that over 28% of the rough grazing present in the United Kingdom in 1945 had been ploughed and reseeded by 1990 (Sinclair, 1992), and in Ireland improved permanent pasture has been advancing at the expense of rough grazing at the rate of 4000 hectares per year since the mid-1980s. In a study conducted for the British government's Department of the Environment, Parry *et al.* (1981) discovered that more than 10% of primary moorland had been lost since the late 1950s, most of it inside the national parks.

Southern member states of the EU have experienced an almost equivalent process of livestock intensification taking place on the best land, linked to the marginalization or slow decline of traditional farming practices elsewhere. In 1986 accession of Portugal and Spain to what was then the European Community proved to be the signal for rapid development of the farm sector, with an often dramatic transformation of farming systems and structures. According to Extezarreta and Viladomiu (1989), an expansion of irrigation in the open Mediterranean landscapes of central Spain and Portugal permitted a general intensification of arable production from this time onwards. Certainly dryland production in Spain is now significantly more intensive than it once was, and monocultural systems are steadily displacing the traditional rich mosaic of arable crops, vines, olives and other permanent crops that offer such an ideal habitat for steppeland birds. Other environmentally valuable permanent crops, such as the traditional olive groves which survive in Greece, Italy, Spain and Portugal, are increasingly neglected or even abandoned.

According to Naveso (1993), traditional olive production in Spain is already economically marginal and liable to be replaced in many areas with a more intensive crop. Portuguese *montados* and Spanish *dehasas*, intricate systems of farming which involve grazing within open woodland and between scattered individual trees (usually cork and holm oaks) together with small-scale arable production, continue to be major depositories of biodiversity in southern farmed landscapes. But intensification threatens here too. Baldock *et al.* (1993) report a steady erosion due to neglect and conversion to more specialized arable use, often facilitated by irrigation. Significant declines in the area of *dehesa* have already been reported in Extremadura and western Andulucia.

Soil erosion was of course not unknown in the United States before 1970. But government programmes were supposed to have dealt with it. Now evidence was being published, notably under the auspices of the National Resources Inventory carried out by USDA, showing a significant upswing in rates of soil loss since the early 1970s. By 1980 an estimated 5.4 billion tons of soil were being eroded each year, much of it from cropland. Meanwhile, evidence was being accumulated by researchers such as Pierre Crosson at Resources for the Future suggesting that off-site damage resulting from the sedimentation of watercourses and from pollution far outstripped those direct on-farm costs due to a loss in soil productivity. A 1985 Conservation Foundation report estimated that damage from sedimentation and pollution might total as much as $13 billion annually; in a later study, USDA put off-site damage costs at $5–18 billion annually (General Accounting Office, 1989).

Commentators began asking, Why, after 50 years of government-financed soil conservation, should there still be a soil erosion problem? The immediate answer was that the ACP itself had not been effective in encouraging soil conservation on farms, an official assessment of the time revealing poor targeting of erosion-prone land and a more or less indiscriminate allocation of conservation monies (General Accounting Office, 1989). But the more profound explanation, on which agri-environmentalists quickly seized, was that environmental effects had not been considered when farmers were being encouraged through the commodity programmes to 'plant fence-row to fence-row'. The result was an almost oceanic change in land use during the middle 1970s, which swept away many conservation practices and brought about a pattern of cropping that was bound to maximize the erosion threat. According to Crosson and Stout (1983), over 13 million hectares were converted to crops between 1972 and 1975, much of it very vulnerable to erosion. By 1981 the area of cropland was greater than at any time in the nation's history.

Indeed, as analysts like Runge (1994) have since shown, the operation of farm policy has determined the intensity of land use as well as the pattern of cropping in US agriculture in ways which greatly exacerbate the environmental impact. On the one hand, farmers managing the two-thirds of cropland enrolled in government programmes have been encouraged to plant more 'programme crops' such as cotton, corn and soya beans, all inherently more erosive and chemically dependent than the crops they often replace. Meanwhile, under the deficiency payments system which operates in the United States, farmers have a strong disincentive to rotate these crops with less erosive ones such as grass or alfalfa because to do so would reduce their eligibility for commodity payments that are calculated with reference to a 'base acreage' of programme crops. On the other hand, USDA's annual acreage reduction programmes actually require farmers to take some cropland out of production in order to continue receiving these payments. The effect here is to encourage a more intensive use of the land which is cropped, given the high product prices which set-aside allows policy makers to maintain, for as Hertel (1990, p. 164) reports, 'the choice of acreage controls is the choice of a high yield/high input system of agricultural production . . . with implications for non-point source chemical pollution and groundwater quality'.

Many critics of the European Community's Common Agricultural Policy (CAP) were by this time similarly persuaded that there was a farm policy explanation for many of the environmental changes being observed in the European countryside. Meanwhile, as Cheshire (1985, p. 14) points out, the entry of member states into CAP meant that 'farmers found themselves operating in an economic environment offering price levels previously undreamt of. Given the rapid acceleration in agricultural change from the 1960s onwards, this is at least consistent with the view that agricultural support plays a significant role in the process of environmental change.' In the short term, the high price guarantees offered to farmers under the CAP encouraged an immediate intensification of production as producers strove to increase output by applying more bought inputs like fertilizers and pesticides to every hectare of land in production and by stepping up stocking rates. As in the United States, a 'coupled' system of agricultural support, linking the subsidy received to the amount produced, gave farmers a powerful incentive to expand output by adopting high input–high output farming systems.

Recent OECD research reported by Rae (1993) has demonstrated that countries with the highest levels of support tend to be the most intensively farmed, with a positive relationship between the intensity of government assistance and the rate of input use. Over the longer term it appears that, once they became convinced that price support was likely to continue, farmers used the short-run profits created by higher prices to re-equip their farms and restructure their farming enterprises, bringing about the specialization of production and the reclamation and improvement of land already described. Very soon, however, they also faced rising land values and rents. As farming became more profitable and more people wanted to become farmers and those already farming wanted to expand, a rising demand for land met a fixed supply, so rents rose along with land values. The resulting inflation of land values has had its own environmental effects, increasing the opportunity cost of uncultivated or unimproved land and providing a spur to land improvement, reclamation and the progressive loss of seminatural vegetation and landscape features which this involves. Cheshire (1985, p. 15) was expressing the consensus view when he wrote that 'the problem is not one of ill will or ignorance but of a system which systematically established financial incentives to erode the countryside, offers no rewards to offset market failure and increases the penalties imposed on farmers who may want to farm in a way which enhances and enriches the rural environment'.

THE INVENTION OF AGRI-ENVIRONMENTAL POLICY

This proved to be an important connection because, as Weale (1992) has commented, environmental policy reform requires both the identification of a problem and the emergence of a consensus about what is causing it. By 1984 both conditions had been met and what might be termed an environmental critique of agricultural policy had gained wide acceptance, even within the farming community. There was apparently equally strong agreement about the need to make agricultural policy more environmentally sensitive and to alter the balance between conservation spending and price support if conservation policies and programmes were to have any chance of success. In the event, progress towards greener farm policies has been slow, a reflection no doubt of the entrenched and culturally embedded nature of agricultural support. By seeking to reform the CAP, British conservationists and later other Europeans were challenging the operation of a policy which, in its basic design, encoded deep assumptions about the European countryside and how it should be managed. They were also challenging a set of policy entitlements which historically had been fiercely defended by a politically astute and institutionally entrenched farm lobby.

Greening has been most dramatic in the United States, where environmental groups had the advantage of quinquennial farm legislation. This created vital windows of opportunity in 1985 and again in 1990. The Food Security Act of 1985 proved particularly productive, for it was being debated at precisely the moment of maximum political saliency for agri-environmental issues. Environmental groups like the American Farmland Trust (AFT) and the Sierra Club lobbied intensively for the inclusion of a conservation title, with the Sierra Club circulating its members with a farm bill alert. The AFT was a particularly important player, marshalling new evidence on the soil erosion problem and putting forward the idea of a conservation reserve which would 'selectively retire highly erodible land'. Nevertheless, the bill's relatively smooth passage (the title became law on 23 December 1985) was largely due to the support given it by powerful farm groups and the gathering perception in Congress that many of its provisions would usefully serve both environmental and production control policy objectives. Certainly the centrepiece of the title, the Conservation Reserve Program (CRP), was presented by groups like the AFT as a multipurpose policy, capable of addressing soil erosion and limiting 'budget exposure' at a time when the administration was embarking on a major round of expenditure cuts to finance tax reductions (American Farmland Trust, 1984).

Under the CRP, farmers would be paid by USDA to take highly erodible land out of cropping and plant it to grass or trees. Sixteen million hecatares were to be targeted in what was hailed as the most significant conservation policy innovation since the 1930s. This was paired with a provision for 'conservation compliance' which stipulated that farmers who wished to retain eligibility for commodity benefits (including deficiency payments but also federal crop insurance and disaster relief payments) must have a conservation plan approved for their farms and implemented by 1 January 1995 at the latest. Under the sod and swampbuster provisions, meanwhile, farmers who ploughed vulnerable grassland or drained wetland after a given date and without a conservation plan faced withdrawal of all farm support and risked further legal penalties. The background to these latter provisions was the long-standing debate about programme consistency, conservation compliance being seen by its supporters as a way of improving the internal consistency of commodity and soil conservation programmes. Its potentially wide application to the 80% of farmers then enrolled in commodity programmes made this mechanism especially attractive to environmental groups. Farm groups were also not unenthusiastic, regarding compliance as a useful way to shore up the legitimacy of farm support payments at a time of growing public criticism.

Lobbyists in the United Kingdom and the rest of the EU did not enjoy the political advantages of regular farm legislation. They did, however, exploit the opportunity created by a review in 1983 and 1984 of the EC Structures Directives (governing policy on investment grants and direct income payments to farmers) to press for the introduction of environmental management payments. Proposals to amend the Less Favoured Areas Directive were scrutinized with particular care. This was traditionally used by agriculture departments to allocate income supports to farmers in areas subject to permanent natural handicaps; conservation groups focused on a proposal to allow payments also to be given to farmers in 'small areas affected by specific handicaps and in which farming must be continued, if necessary subject to certain conditions, in order to ensure the conservation of the environment, to maintain the countryside'.

In evidence to a House of Lords select committee, the Countryside Commission and voluntary bodies claimed this opened the door to monies from the agricultural budget being used to subsidize environmental management on farms. It was a beguiling argument, initially rejected on legal grounds by UK agriculture departments and the Department of the Environment, but later taken up by the agriculture minister and energetically promoted in Brussels. The resulting article of the 1985 Structures Regulations permitted agriculture departments to designate Environmentally Sensitive Areas (ESAs) where 'the maintenance or adoption of particular agricultural methods is likely to facilitate conservation, enhancement or protection'. Ten-year management agreements could now be offered to farmers within these ring-fenced areas in return for annual hectarage payments.

The UK agriculture departments were the first to implement the new policy, acting quickly to designate seven ESAs in 1986/87 and a further six in 1990. By 1994 there had been some 5600 agreements signed with farmers involving almost 400 000 hectares of land (MAFF, 1995) – a substantial addition to the conservation estate compared to the 1.7 million hectares in total enrolled in management agreements on conservation sites up to that date. Other member states followed, so that by 1992 there were agri-environmental programmes in place in the Netherlands, Germany, Denmark and France. Later, in 1992, a new agreement led to the setting up of a more fully-fledged agri-environmental policy which was binding on all member states and extended payment to a wider range of conservation activities and investments (Table 5.1, overleaf). Agri-environmental policy had become a permanent feature of the agricultural policy scene.

TOWARDS GREENER FARM POLICIES?

On one level it would be hard to exaggerate the significance of these developments in policy terms, particularly in the United States, where omnibus agricultural and environmental legislation has brought about some very large policy interventions indeed. By common consent, the CRP has greatly extended the soil conservation effort by targeting 'fragile land' and paying farmers to take it out of production. As Table 5.2 (page 95) shows, from the first sign-ups, farmers have been very willing to enrol land in the programme and at its close in 1992 the reserve contained 14.5 million hectares, accounting for almost 11% of total US cropland. According to Osborne and Miranowski (1994), the effect has been to reduce erosion by 700 million tons or more per year, or 22% of the total amount of soil lost each year on

Table 5.1   Land enrolled in the US Conservation Reserve Program

Sign-up period	Number of contracts (thousand)	Total acres enrolled (million)
March 1986	9.4	0.75
May 1986	21.5	2.77
August 1986	34.0	4.70
February 1987	88.0	9.48
July 1987	43.7	4.44
February 1988	42.7	3.38
July 1988	30.4	2.60
February 1989	28.8	2.46
August 1989	34.8	3.33
May 1991	8.6	0.48
July 1991	14.7	1.00
June 1992	18.4	1.03
Total	375.0	36.42

Source: Osborne (1993).

US farms. Conservation compliance has also been hailed as an important policy departure, for whereas CRP relies on what Vail *et al.* (1994) call 'the producer compensation principle' – persuading farmers to do something by offering them government subsidies – conservation compliance threatens the withdrawal of commodity payments if conservation conditions are not met, a decidedly more radical proposition. An estimated annual erosion saving of over 380 million tons is attributed to this provision. Indeed, Dickason and Magleby (1993) estimate that by 1995 the CRP and conservation compliance provisions combined will have achieved twice the amount of erosion control accomplished by all conservation practices previously implemented on US cropland since the 1930s.

The EU's Agri-Environmental Programme (AEP) is a more modest and experimental affair, the incremental product of a slower-moving policy process. Nevertheless, it too has been welcomed as an important policy innovation which is redefining the way agricultural landscapes and farmland habitats are valued, managed and conserved throughout the EU. Supporters point to the injection of new money into countryside protection and agri-environmental pollution control, which the commission's 2.16 billion ecu contribution will make over the first five years of the programme and to the significant fact that all future funding will be drawn from the guarantee section of the CAP rather than, as previously, the much smaller guidance fund. In nature protection terms, it shifts the conservation effort decisively away from conservation sites in favour of more extensive tracts of countryside and, while the British Ministry of Agriculture has resisted calls to make ESA agreements available throughout the wider countryside, it has effectively doubled the size of the conservation estate through its agri-environmental programmes.

Adams (1986) comments that there has been a long-standing debate in conservation circles about the need for wider tracts of countryside to be reserved and

Table 5.2 Agri-environmental schemes implemented in EU member states[a]

Element of Regulation 2078/92	B	Dk	D	Gr	E	F	IR	I	L	NL	A	P	Fin	S	UK
Reduction of chemical inputs and pollution control	✓	✓	✓	✓	✓	✓	✓	✓	✓	✓	✓	✓	✓	✓	✓
Organic farming	✓	✓	✓	✓	✓	✓	✓	✓	✓	✓	✓	✓	✓	✓	✓
Extensification	✓	✓	✓	✓	✓	✓	✓	✓	✓	✓	✓	✓	✓	✓	✓
Convert arable into grassland	✓	✓	✓	✗	✓	✓	✗	✓	✓	✗	✗	✓	✗	✓	✓
Reduction of livestock density	✓	✗	✓	✗	✓	✓	✗	✓	✓	✗	✓	✓	✗	✗	✓
Environmental practices	✓	✓	✓	✓	✓	✓	✓	✓	✓	✓	✓	✓	✓	✓	✓
Landscape and countryside management	✓	✗	✓	✗	✓	✓	✓	✓	✓	✓	✓	✓	✓	✓	✓
Rearing of local breeds in danger of extinction	✗	✗	✓	✗	✓	✗	✗	✓	✗	✗	✓	✓	✓	✓	✗
Upkeep of abandoned farmland or woodland	✗	✗	✓	✓	✓	✗	✓	✓	✓	✗	✓	✓	✓	✓	✗
Twenty-year set-aside	✗	✓	✓	✓	✓	✓	✓	✓	✗	✗	✓	✗	✓	✗	✓
Public access and leisure	✗	✗	✗	✗	✓	✓	✓	✓	✗	✓	✗	✓	✗	✗	✓
Training and demonstration projects	✓	✗	✓	✓	✓	✓	✓	✓	✗	✓	✓	✓	✓	✓	✗

[a] Elements only have a tick if they are a main objective of a scheme.
Sources: de Putter (1995) and various STAR documents.

Table 5.3 Distribution of CRP enrolments, 1985–1988

Region	Percent of soil loss on eligible area from			Percent of enrolled ha	Percent of soil saved on enrolled area
	Wind	Water	Total		
Mountain states	13	3	16	20	19
Northern Plains states	5	7	12	25	20
Southern Plains states	22	3	25	16	26
Subtotal	**40**	**13**	**53**	**61**	**66**
Northeast states		3	3	1	–
Lake states	1	3	4	8	7
Corn belt states	1	25	26	14	13
Appalachian states		7	7	3	4
Southeast states		2	2	5	4
Delta states		2	2	3	3
Pacific states	1	3	3	5	3
Subtotal	**3**	**44**	**48**	**39**	**34**
Total	**43**	**57**	**100**	**100**	**100**

Source: GAO (1989).

protected. ESAs do not quite achieve this, but they do allow for the designation of areas that are large enough to be seen as distinct farmed landscapes. The wider use of environmental contracts made possible by AEP means this is arguably one of the biggest experiments in the use of incentives to influence and reward conservation behaviour. More than that, it promotes conservation on a European scale, albeit through a suitably decentralized set of policy arrangements.

But doubts remain, about the environmental effectiveness of the measures them-selves, their significance within the wider scheme of things and the advisability of linking conservation so directly to a particular set of sectoral policies. In the United States, criticism of the CRP mounted steadily once it became clear that USDA had failed to enrol either the most erodible or environmentally vulnerable land in the first years of the policy's operation. A report from the General Accounting Office (1989), 'CRP Could be Less Costly and More Effective', soon became the verdict of many commentators. Revealing a strong bias in favour of wind-eroding rather than water-eroding land, with a heavy concentration of enrolment in the Corn Belt, Delta and Lakes states (Table 5.3), the report concluded that the reserve was much less effective than it might have been in reducing off-site damage.

USDA's objective, it appeared, was to enrol as much productive land as pos-sible, as quickly as possible during these early sign-ups in order to maximize the production control effect of the policy. Now this was perfectly consistent with the legislation, which actually gives priority to supply control over all other goals, a reflection in turn of the compromises agri-environmentalists had to make to get the conservation title through. Indeed, it was the CRP's 'magic bullet' property – its apparent ability to reduce agricultural output, support farmers' incomes and promote soil conservation – which so appealed to otherwise sceptical members of Congress. As Swanson (1992, p. 112) says, 'by pulling highly erodible land out of cropping Congress was able to reduce total production and pay 40 million dollars to farmland areas while seeming to address serious problems of soil erosion'.

Policy makers were soon persuaded that the CRP's conservation mission could only be rescued through better targeting of subsidies at the environmentally most vulnerable land. Targeting preoccupied environmental groups in the run-up to the 1990 Farm Bill; and under the Food, Agriculture, Conservation and Trade Act, which became law later that year, provision was made to rebalance the CRP by tightening the enrolment criteria to concentrate on conservation priority areas. A new Wetlands Reserve Programme (WRP) designed to recreate wetland habitat across the United States was also set up. This targeted up to 400 000 hectares and required farmers to agree to have permanent easements written into the title deeds to their land in return for payment. Clearly, the conservation emphasis of the policy had been greatly strengthened. But it was hard to escape the impression that agri-environmental innovations were not always what they seemed.

European commentators were rapidly coming to the same conclusion. The prominence given to the aim of 'ensuring an adequate income for farmers' in the initial regulation was inevitable given the political bargain between environmental and farm groups which lay behind the ESA initiative. But a later amending regulation – justifying the adoption of environmentally benign practices on farms in order 'to contribute to the adaptation and the guidance of agricultural production, accommodation to market needs and having regard to the income losses resulting from this' (CEC, 1987) – caused some to question the likely efficiency of a policy with so many (potentially contradictory) objectives (e.g. Baldock and Beaufoy, 1992). It is a reasonable assumption that, from the perspective of many member states, agri-environmental schemes are little more than disguised income payments which have been introduced at a time when direct income aids are becoming increasingly difficult to justify in political terms. At the very least, it is difficult to judge the real environmental benefits of many 'extensification' schemes that have been introduced, and there is concern about the poor design and justification of some schemes now being set up in southern member states (Wilkinson *et al.*, 1994).

For countries like the United Kingdom, Germany, Netherlands and Denmark, with a longer history of involvement in AEP programmes, a distinction is increasingly being drawn between the ability of agriculture departments to bring about a change in policy outcomes – setting up the schemes and enrolling sufficient farmers to make a difference – and policy results: the short- and long-run environmental benefits that are actually produced on farms and in fields. Questions are being asked about the sort of improvements in environmental and nature protection which conservation by contract is able to bring about. One aspect of this is the extent to which measures bring about changes in the attitudes and behaviour of farmers which outlast the schemes themselves. It could be argued that until they can be shown to have such effects, the long-term political sustainability of agri-environmental policy will continue to be in doubt.

In any event, it is still far from clear that the greening process has altered the structure of agricultural support itself in any very significant way. Critics argue that, however robust agri-environmental programmes may be, little lasting environmental improvement will be achieved so long as the bulk of farm subsidies continue to be delivered through the apparatus of commodity price support. US agricultural economists like Ford Runge (Runge, 1994) have argued for some time that commodity support cuts across, and ultimately cancels out, the government's conservation effort, so government efforts achieve less and cost more than might otherwise be the case. In an obvious sense, high price support increases the profit farmers can make by having their land under crops and thus inflates the price USDA must pay to persuade them to put land in CRP-type programmes. More technically, under the American deficiency payment system, commodity price support is calculated in terms of a farmer's 'base acreage' of crops grown in a given year. Putting land into the CRP erodes this base and thus eligibility for lucrative deficiency payments. Again, the effect is to dissuade farmers from putting land into the CRP or, if they do, to make it more likely they will demand high payments.

Similar contradictions surround agri-environmental policy under the CAP. Relatively slim resources have been allocated to the AEP in the initial five years of the programme (equivalent to just 2% of total farm spending). This encourages the view that the greening process has been largely cosmetic, at least to date, and has left intact the powerful incentives for agricultural intensification. It is certainly true that west European farmers continue to receive high price guarantees under the CAP (agricultural spending exceeded 68 billion ecu in 1994); policy makers were slow to reduce price support despite pressure to do so from international trading partners. If anything, the CAP has become more Byzantine in its complexity following the much vaunted MacSharry agreement of 1992, which implemented modest price cuts but offset them by also setting up various compensation schemes. One of them, the Arable Areas Payment Scheme, is conditional on farmers setting aside some of their arable land (Chapter 4).

The policy is already bringing about some of the most dramatic land-use changes since the 1970s, with over 6.2 million hectares being removed from production throughout the EU in the 1993/94 marketing year. But few agri-environmentalists see this as a welcome or wise development, aware as they are that set-aside allows policy makers to operate a high price policy while avoiding the problems of surplus production. In the United Kingdom, the Ministry of Agriculture's determination to apply a green gloss to what is essentially a supply control policy, has meant that farmers can choose to set aside the same patch of land for up to six years and apply for additional payments in return for managing the retired land in more environmentally sensitive ways than simply fallowing it.

However, concern remains that the set-aside scheme will 'crowd out' AEP measures in precisely the same way that USDA's ARPs crowd out the CRP. Farmers required to set aside the stipulated 10% of their arable area are going to be increasingly reluctant to find yet more land for AEP schemes; or if they do, they are going to require higher payments to do so. With policy makers torn between ensuring that set-aside is efficient in controlling production (hence insisting that set-aside land is rotated around the farm) and wishing to secure environmental benefits (by conceding that the same piece of land can be set aside for longer periods after all), one can only agree with Winters (1987) that the tendency 'to do something' is always greater than the imperative to 'undo something' in the agricultural policy field. The result is a 'costly tangle of potentially inconsistent instruments' (Winters, 1987, p. 310).

In their efforts to capitalize on this new policy situation, and the partial substitution of direct payments for market intervention – the main consequence of the MacSharry reforms – environmental groups, especially in the United Kingdom, have begun to press for US-style conservation compliance to be applied to the CAP. Cross-compliance of sorts already exists with the set-aside condition attached to the arable area payments, but supporters of conservation compliance (e.g. Dixon and Taylor, 1990) would like to see it much more directly geared to environmental improvement, with farmers being required to put land into agri-environmental schemes in order to qualify for production premiums or compensation payments of any sort. Their argument is that compliance offers policy makers an opportunity to influence rural land use throughout the countryside, not just on the land which happens to be volunteered for an agri-environmental scheme.

The majority of arable farmers are now in receipt of arable payments, for instance, and livestock payments are the mainstay of many grassland farms, particularly in upland areas, where the system of headage payments and other subsidies available under the sheepmeat regime made up a high proportion of total farm receipts. This gives policy makers potentially enormous leverage. As a concept, conservation compliance is very much in step with ambitious new thinking about habitat and landscape protection, which the agri-environmental reforms of the 1980s have helped to bring about; it promotes the idea that conservation is best 'produced' jointly with food throughout the countryside by suitable adjustments to farming practice and seeks to influence land use on as broad a front as possible. Politically, it has already gained the support of the farm lobby, at least in the United Kingdom (e.g. National Farmers' Union, 1995; Country Landowners' Association, 1994), who can see advantages in a reform which would enable them to proclaim the public goods farmers produce in return for government support.

US agri-environmentalists will be familiar with many of these arguments, because conservation compliance has been built into US agricultural support since 1985. It has not been an unqualified success. First, USDA has been reluctant to apply the powers which the 1985 act bestows, penalizing very few landowners who have ploughed erodible grassland or drained wetland over the period. Secondly, and perhaps most decisively, it has been discovered that the match between farmers dependent on commodity programmes and those with vulnerable land is far from ideal. As Reichelderfer (1985) predicted, efforts to increase the consistency of USDA commodity and conservation programmes in this way are not necessarily the most efficient way to tackle the soil erosion problem. This is evidence to show how the location of vulnerable wetland habitats is far from coincident with the geographical distribution of government support. Moreover, the effectiveness of conservation compliance has already begun to decline as commodity price support is scaled down and more farmers come out of government programmes (see below). All of which should give European conservationists pause, if only to contemplate the political drawbacks of a policy arrangement with a limited shelf-life. It is at best a blunt instrument, a pragmatic way to build conservation conditions into existing state aids but one which declines in effectiveness once those payments are scaled down or more carefully targeted.

CONSERVATION IN A PERIOD OF FARMING CONTRACTION

This raises a larger question: are further attempts to 'green' existing systems of agricultural support either feasible or desirable in the long term; or should conservationists be preparing for a future in which government support to agriculture no longer exists, at least in its present form? Brubaker and Castle (1982, p. 312) were warning US groups some time ago that 'it may be unwise to tie long term concerns for soil erosion to an array of agricultural policies that may not deserve continued support on their own merits'. This argument has now taken on added force in the wake of a successful GATT round committed to the eventual elimination of agricultural protectionism. Everything points to a progressive scaling down and decoupling of farm policies throughout developed countries in the years ahead as international reform gathers pace. Already in the United States there is evidence of a sea change in attitudes towards the commodity programmes, and the effect of the 1995 Farm Act will be to fashion an agricultural industry that is less protected and more export-driven than ever before.

European policy makers face mounting pressures at home and abroad to make further inroads into the price guarantees that have traditionally been at the heart of the CAP. EU enlargement to embrace the countries of eastern and central Europe at the turn of the century will put further strain on the policy, greatly expanding productive capacity and therefore the cost of surplus disposal. Meanwhile, the approach to a new GATT round in 2002/03 will create an incentive to decouple the CAP in order to gain negotiating advantage. In any event, the disciplines of a new GATT agreement, when it arrives, will almost certainly require a substantial recasting of farm support throughout the developed world.

The implications for rural land use and for the conservation of soil, water, habitats and landscapes are profound. One result could be the expansion of agri-environmental policies as some of the 'peace dividend' of international agricultural policy disarmament is channelled into programmes that can still be justified on the grounds of public good. There are commentators who fully expect green payments to have become the principal means of support by this stage. Indeed, many farm groups already seem perfectly reconciled to this outcome; the National Farmers' Union (1995, p. 34) in the United Kingdom agrees that 'the decoupling option would provide an acceptable means of directing public support for the provision of specified public goods'. It is now widely accepted that the decoupling of farm support is likely to be a good thing for the environment because it cancels the production subsidies which have for so long worked against agri-environmental interests and it paves the way for a green recoupling of farm support.

In reality, negotiating the transition to a radically different policy regime will not be easy. It is very unlikely that green payments will fill the gap in funding left by conventional support. An issue of political legitimacy is at stake; although tax-payers may indeed be more willing to pay farmers to protect the countryside and conserve the soil than to subsidize farm incomes *per se*, they are unlikely to be persuaded to meet such a high cost. Methodologies to prescribe and evaluate envir-onmental improvement on farms are still poorly developed and policy makers are some way from developing a system of paying by results that would arguably be required for highly visible and 'transparent' green payments to be politically defens-ible on a large scale. This funding gap means that many more land-use changes will be driven by the market than determined by agri-environmental policy, and pro-bably not always in the direction that conservationists would like. Conservationists and policy makers will need to confront the social and environmental repercussions of the substantial net withdrawal of government support which green recoupling necessarily entails.

It is easier to regard these outcomes as beneficial in a US context where, other things being equal, the progressive extensification of production and shift in crop-ping patterns expected to accompany the ending of commodity support will ease soil erosion, improve water quality and reduce the incentive for wetland conversion (Reichelderfer, 1990). Here it is a safe assumption that federal government will continue to fund the set-piece conservation programmes, though Osborne (1993) predicts a swing away from long-term land diversion programmes, like CRP, back towards traditional cost-sharing arrangements which subsidize the installation of specific conservation practices on farms. Funding for CRP and WRP, though still substantial in absolute terms, is already declining steeply as the first 10-year con-tracts expire, and Osborne expects that no more than $1 billion annually will be committed between 1995 and 2000 compared to over $2 billion in the late 1980s. Current public concern about what will happen to CRP land when contracts expire underlines the temporary nature of the environmental improvement which land retirement is able to bring about and is leading to calls for a more selective and permanent diversion of land through a wider use of conservation easements (restrictions written into the title deeds to land) rather than 10-year contracts.

Policy makers seem less persuaded than they once were of the chief merit of land retirement programmes – their ability to bring about definite land-use changes for a specified duration – and more anxious to set up broader stewardship-type schemes, which appear to promise changes in farmer behaviour without continuing government payments. But as European agri-environmentalists are discovering, there is a danger that broader-brush schemes have a greater chance of capture by farm groups for farm income support purposes. Far from heralding a move towards policies with clearer-cut conservation motives, the decoupling process could make the invention of hybrid schemes more likely, especially if farm groups continue to hold the policy reins and conservation groups are unable to come up with measurable policy targets.

For European agri-environmentalists, an agricultural industry more exposed to the forces of international competition than for decades past is a mixed prospect. Optimists (e.g. MAFF, 1995) predict that dismantling price guarantees will work in the same direction as existing agri-environmental measures, bringing about a less intensive use of land, slowing down land reclamation and pushing more farmers into the schemes themselves as returns from agriculture fall relative to the conservation payments on offer under an expanded AEP. In practice, it is far from certain that complex and spatially uneven restructuring of production and land use precipitated by deep price cuts would everywhere benefit the farmland habitats and landscape mosaics which make up the European conservation resource.

More than in the United States, the phasing out of price support, particularly if followed by the dismantling of transitional compensation schemes, will generate enormous pressures for agricultural restructuring, eliminating many marginal farmers, some of them operating high natural value farming systems, and accelerating the trend towards fewer, larger holdings. Researchers such as Harvey *et al.* (1986) argue that arable production will become further concentrated on the best land whereas mixed farming will continue its long-term decline. Increasingly, it will be shifts in land use out of agriculture and into forestry and industrial crops, especially on the extensive margins of production, rather than 'smooth' changes in farming practice, which will become the focus of conservation concern (Lowe *et al.*, 1995, ch. 6).

Meanwhile, the ending of conservation compliance will limit policy makers' ability to engineer land-use changes on a broad front, and with limits being reached to expanded AEP funding, combinations of public and private funds (including direct ownership of land by conservation groups) may be needed to satisfy growing public demand for the conservation and recreation of locally distinct, as against nationally important, farmed landscapes. This hybridization of the conservation effort notwithstanding, funding for AEP measures is likely to increase, though their more selective deployment necessary to maximize environmental value for money will demand a stricter prioritization of objectives and a more imaginative approach to the conservation of agricultural landscapes. As Green (1995) comments, the idea that farming is needed everywhere to maintain valued environments will need to be re-examined, with the creation of wilderness areas becoming a distinct possibility. This will be controversial with many member states, particularly where the cultural desirability of a 'landscape with figures' is more firmly entrenched. A tightly drawn EU AEP is difficult to imagine in these circumstances, and while landscape and habitat conservation will increasingly be conceived and publicly justified at an EU scale, its achievement in practice will continue to be a largely national concern.

CONCLUSIONS

The invention of agri-environmental policy in the United States, the European Union and many other developed countries is surely one of the most impressive achievements in the environmental field of recent years. Partly a reaction to the sweeping changes in farming practice of the 1960s, 1970s and 1980s, the greening of agricultural policy reflects a profound public reassessment of farmers and the relationship between agriculture and the environment. Rather than resort to government regulation and control, however, the approach followed has been to subsidize farmers so they can better fulfil their stewardship obligations. In the United States, initiatives like the CRP have brought about land-use change on an unprecedented scale and a very significant degree of new investment in conservation capital; in the European Union, a large experiment is effectively in progress involving paying out of the farm budget for farmers to conserve agricultural landscapes as well as to deintensify production in order to protect the resources and reduce pollution.

Despite considerable environmental and institutional differences, there are impressive parallels between the American and European stories. In both cases, the greening process has been a rather fortuitous affair, the product of a series of political bargains between agri-environmentalists and still very powerful farm groups. Closer examination reveals how the policies themselves are not always what they seem. Little agri-environmental expenditure is for strictly environmental purposes, and too many schemes are barely more than disguised income support schemes for farmers. A fundamental rethink of what government support to agriculture is for, and who desrves to receive it, has yet to take place. With the international liberalization of agricultural trade looking increasingly inevitable, this cannot long be delayed.

Tactically, there are various opinions about how best to secure conservation of soil, water, habitats and landscapes in a period of international farm policy reform. Conservation compliance is an idea that has gained much support in both the United States and the European Union. A major drawback is that, as a strategy, it assumes income and producer aids will be permanent features of the policy scene, when under GATT auspices, international reformers are working towards their eventual elimination. By comparison, green recoupling – working towards a system in which all support to farmers is delivered through environmental schemes – seems much more in step with long-term policy trends, though it poses formidable problems in terms of policy design. It also implies the creation of a largely market-driven agricultural industry and the enormous restructuring of farms and land use this will involve. Having spent much of the past decade questioning the doctrine of agri-environmental compatibility, conservationists may now need to plan how best to manage agricultural adjustment in yet another period of change.

REFERENCES

Adams, W.M. (1986) *Nature's place: conservation sites and countryside change.* Allen and Unwin, London.

American Farmland Trust (1984) *Soil conservation in America, What do we have to lose?* AFT, Washington DC.

Baldock, D. and Beaufoy, G. (1992) *Green or mean? Assessing the environmental value of the CAP reform 'accompanying measures'.* Council for The Protection of Rural England, London.

Baldock, D., Beaufoy, G., Bennett, G. and Clark, J. (1993) *Nature conservation and new directions in the EC CAP*. Institute for European Environmental Policy, London.

Barr C., Bunce, R., Clarke, R., Fuller, R., Furse, M., Gillespie, M., Groom, G., Hallam, C., Hornung, M., Howard, D. and Ness, M. (1993) *Countryside survey 1990. Main Report*, Department of the Environment, London.

Batie, S. (1988) Agriculture as the problem: new agendas and opportunities. *Southern Journal of Agricultural Economics*, **20**, 1–12.

Brubaker, S. and Castle, E. (1982) Attention policies and strategies to achieve soil conservation. In Halcrow, H., Heady, E. and Cotner, M. (eds) *Soil conservation policies, institutions and incentives*. Soil Conservation Society of America, Andency, pp. 302–19.

CEC (Commission of the European Communities) (1987) *Agriculture and environment: management agreements in 4 countries of the European Communities*. CEC, Brussels.

Cheshire, P. (1985) The environmental implications of European agricultural support policies. In Baldock, D. and Condor, D. (eds) *Can the CAP fit the environment*. IEEP/CPRE/WWF, London, pp. 9–18.

Country Landowners' Association (1994) *Focus on the CAP*. CLA, London.

Crosson, P.R. and Stout, A.T. (1983) *Competitiveness and productivity effects of cropland erosion in the United States*. Resources for the Future, Washington DC.

de Putter, J. (1995) *The greening of Europe's agricultural policy: the agri-environmental regulation of the MacSharry reform*. Ministry of Agriculture, Netherlands.

Dickason, C. and Magleby, R. (1993) *Erosion reduction benefits for US soil conservation*. Working Paper, US Department of Agriculture, Economic Research Service, Washington DC.

Dixon, J. and Taylor, J. (1990) *Agriculture and environment: towards integration*. Royal Society for the Protection of Birds, Sandy.

Dower, J. (1945) *Conservation of nature in England and Wales*, Report of the Wild Life Conservation Special Committee. HMSO, London.

Extezarreta, M. and Viladomiu, L. (1989) The restructuring of Spanish agriculture and Spain's accession to the EEC. In Goodman, D. and Redclift, M. (eds) *The international farm crisis*. Macmillan, pp. 135–55.

Felton, M. (1993) Achieving nature conservation objectives: problems and opportunities with economics. *Journal of Environmental Planning and Management*, **36**(1), 23–31.

General Accounting Office (1989) *CRP could be less costly and more effective*. GAO, Washington DC.

Green, B. (1995) Plenty and wilderness? Creating a new countryside. *ECOS*, **16**(2), 3–9.

Harvey, D., Barr, C., Bell, M., Bunce, R., Edwards, D., Errington, A., Hollans, J., McClintock, J., Thompson, A. and Tranter, R. (1986) *Countryside implications for England and Wales of possible changes in the Common Agricultural Policy*. Centre for Agricultural Strategy, University of Reading.

Hays, S.P. (1959) *Conservation and the gospel of efficiency: the progressive conservation movement, 1890–1920*. Harvard University Press, Cambridge MA.

Hays, S.P. (1987) *Beauty, health and permanence: environmental politics in the United States, 1955–1985*. Cambridge University Press, Cambridge.

Hertel, T.W. (1990) Ten truths about supply control. In Allen, K. (ed) *Agricultural policies in a new decade*. Resources for the Future, Washington DC, 153–69.

Huxley, J. (1947) *Conservation of nature in England and Wales*, Cmnd 7122. HMSO, London.

Lowe, P., Ward, N., Ward, J. and Murdoch, J. (1995) *Countryside prospects, 1995–2000: some future trends*. Centre for Rural Economy Research Report, Newcastle.

MAFF (1995) *European agriculture: the case for radical reform*. MAFF, London.

Meeus, J., Wijermans, M. and Vrom, M. (1990) Agricultural landscapes in Europe and their transformation. *Landscape and Urban Planning*, **18**, 189–352.

Naveso, M. (1993) Estepar, aves y agricultura. *La Garcilla*. SEO/Birdlife International, Madrid.

Nature Conservancy Council (1979) *Nature conservation and agricultural change*. NCC, Peterborough.

National Farmers' Union (1995) *Taking real choices forward*. NFU, London.

Osborne, T. (1993) The conservation research program: status, future and policy options. *Journal of Soil and Water Conservation*, **Jul/Aug**, 272–79.

Osborne, T. and Miranowski, J. (1994) Lessons from US policies for soil erosion control. In OECD (eds) *Agriculture and the environment in the transition to a market economy*. Organization for Economic Cooperation and Development, Paris, 221–30.

Parry, M., Bruce, A. and Harkness, C. (1981) The plight of British moorlands. *New Scientist*, **90**, 550–52.

Rae, J. (1993) Agriculture and the environment in the OECD. In Williamson, C. (ed) *Agriculture, the environment and trade*. International Policy Council on Agriculture and Trade, Brussels, pp. 82–114.

Reichelderfer, K.H. (1985) *Do USDA farm program participants contribute to soil erosion?*, USDA Economic Research Service, Report 532, Washington DC.

Reichelderfer, K.H. (1990) Environmental protection and agricultural support: are trade-offs necessary? In Allen, K. (ed) *Agricultural policies in a new decade*. Resources for the Future, Washington DC, pp. 201–18.

Reichelderfer, K.H. (1992) Land stewards or polluters? The treatment of farmers in the evolution of environmental and agricultural policy. In Swanson, L. and Clearfield, F. (eds) *Farming and the environment*. Soil and Water Conservation Society, Ankeny IA.

Runge, C.F. (1994) Environmental incentives for agriculture: carrots, sticks, and conditionality. Unpublished paper prepared for the Environment Directorate, OECD, Paris.

Sheail, J. (1994) War and the development of nature conservation in Britain. *Journal of Environmental Management*, **44**, 267–83.

Sheail, J. (1995) Nature protection, ecologists and the farming context: a UK historical context. *Journal of Rural Studies*, **11**(1), 79–88.

Shoard, M. (1984) *The theft of the countryside*. Temple-Smith, London.

Sinclair, G. (1992) *The lost land: land use change in England 1945–90*, Council for the Protection of Rural England, London.

Swanson, L. (1992) Agro-environmentalism: the political economy of soil erosion in the USA. In Harper, S. (ed) *The greening of rural policy*. Belhaven Press, London, pp. 99–117.

Vail, D., Hansund, K. and Drake, L. (1994) *The greening of agricultural policy in industrial countries*. Cornell University Press, Ithaca NY.

Weale, A. (1992) *The new politics of pollution*. Manchester University Press, Manchester.

Wilkinson, A., Bernstein, N., Delorme, H., Heneriks, G., Berkhout, P., Meester, G. and Nedergaard, P. (1994) Renationalisation: an evolving debate. In Kjeldahl, R. and Tracy, M. (eds) *Renationalisation of the CAP*? Institute of Agricultural Economics, Copenhagen.

Winters, L. (1987) The political economy of the agricultural policy of industrial countries. *European Review of Agricultural Economics*, **14**, 285–304.

CHAPTER 6

THE CHANGING ROLE OF FORESTS
Alexander Mather

INTRODUCTION

The role of forests varies enormously within the developed market economies (DMEs). In some, such as Finland and Japan, forests occupy two-thirds or more of the national land area, whereas in others, such as the United Kingdom and Ireland, the proportion is one-tenth or less. In some countries, such as Canada and Sweden, forests play a major part in the national economy, whereas in others their role is relatively insignificant. DMEs include both the leading importers and leading exporters of forest products. Furthermore, the role of the forest in national culture and history varies greatly, not only between the extremes of Scandinavia and north-west Europe, but also more subtly between countries such as Germany and France.

With such variations and contrasts, useful generalization about the the role of forests in rural change is not easily achieved. Nevertheless, certain general trends operate widely. One is that the forest area is increasing; another is that perceptions of forests and of forest values have changed and are changing, to the extent that paradigm shifts have occurred in resource perceptions. As in the wider rural sphere, changes in land use (in the sense of forest area) are linked to changes in rural perceptions and attitudes. This chapter reviews both forest trends. At first sight they seem unrelated, but closer examination reveals complex interaction between changing forest perceptions and increases in forest area. Although diverse national circumstances are acknowledged, space does not permit their detailed evaluation. This being so, two countries – the United Kingdom and Finland – are used as frequent examples which represent contrasting ends of the spectra of forest extents and economic importance. Before the modern trends are examined, however, some of the basic and contrasting characteristics of forests in different parts of the developed world are outlined.

FORESTS IN DMEs

Table 6.1 summarizes a few key features of forests in selected DMEs. Clearly the forest extent and per capita forest endowment vary widely within the developed

Table 6.1 Basic forest characteristics[a]

Country	Forest area as percent of total area[b]	Forest area per capita[b] (ha)	Ownership[c] (%)					Farm and other private holdings	
			State	Other public body	Industry	Farm	Other	Number	Average size
Austria	47	0.5	15.0	3.1		53.1	28.8	227 774	14
Belgium	20.5	0.06	10.3	33.7		1.3	54.7	11 000	3
Denmark	11	0.09	24.0	3.2			72.7	35 700	9
Finland	76.7	4.68	23.7	2.5	8.7	62.8	2.2	426 303	28
France	26.1	0.25	10.1	15.9				3 677 000	3
Germany	30.7	0.13	40.3	16.9				441 856	7
Ireland	6.2	0.12	83.8	0.5			15.7		
Italy	28.4	0.15	7.2	32.8					
Netherlands	9.8	0.02	31.1	16.5		6.0	46.4		
Norway	31.2	2.26	12.0	4.0	4.0		80.0	145 075	46
Portugal	35.8	0.29	2.5	7.1	6.7	83.7		373 669	7
Spain	51.3	0.66	7.7	31.0	4.2		57.1	4 822 541	4
Sweden	68.6	3.27	18.9	9.9	23.4	47.8		248 879	53
Switzerland	29.8	0.18	5.6	62.2		28.8	3.5	251 700	2
UK	9.9	0.04	42.5						
Canada	49.2	17.09	90.6	0.1	1.2	8.1			
USA	32.4	1.18	26.8	1.5	14.6	20.1	37.1		
Japan	67.8	0.2	43.5					234 276	11
Australia	19.3	8.52	27.5	43.8			28.7		
New Zealand	27.9	2.23	74.3	25.7					

[a] Where no figures are shown, data is unavailable.
[b] Forest area includes other wooded land (columns 1 and 2) but ownership data relates to forest only.
[c] Ownership data is for West Germany only.
Source: UNECE/FAO (1992).

world. Equally marked contrasts exist in terms of ownership. Compared with agricultural land, the relative importance of state ownership and other forms of public ownership (e.g. by local authorities) is much greater. Within the private sector, however, there are also wide differences. The statistical base is incomplete and many countries, including the United Kingdom, are unable to provide a detailed breakdown of private forest ownership. Nevertheless, it is clear that ownership patterns in some countries are characterized by large numbers of small units. In the Scandinavian countries and in central Europe, there has traditionally been a close link between farming and forestry; every farm in Finland has an area of forest, and around 40% of the forest properties are combined with agriculture (Varjo and Tietze, 1987; Nordic Council, 1990). In contrast, industrial ownership is usually associated with much larger units, although its aggregate extent is limited.

Different types of ownership often display contrasts in spatial patterns and in management. State ownership in Sweden and Finland is associated with the less productive forests of the North; industrial and farm ownership are more usually found on the more productive and more accessible areas. Almost by definition, industrial ownership is associated with maximum timber production. In Finland the silvicultural quality of management (for timber production) is often highest in company-owned forests (Kuusela and Salminen, 1991). Management among non-industrial private forest (NIPF) owners often differs both in style and in objectives, and within the NIPF sector there may be contrasts between farmers and other owners. Although the primary objective of farmer-owners may be timber production as a source of income, some other NIPF owners are more oriented towards recreation or a pleasant place of residence. As rural society changes, forest management will change with it. The implications of these changes are considered more fully later in the chapter.

EXPANDING FORESTS

A distinctive feature of the DMEs is the fact their forests are generally stable or increasing in extent. In this respect they contrast with the world as a whole, and with the developing world in particular. Up until the nineteenth century, forests in most DMEs were shrinking probably as rapidly as those in most developing countries today. Since then, however, expansion has been the dominant trend in most DME countries. The process of expansion is most clearly apparent in Europe. Long-term statistics on forest extent are notoriously inadequate, but one respected source suggests the European forest area in the late twentieth century was slightly greater than it had been in 1850 (World Resources Institute, 1987). The forest area in Japan stabilized and began to increase a century earlier than in Europe (e.g. Osaka, 1983; Totman, 1986). In the United States, the rapid deforestation of the nineteenth and previous centuries had slowed by the early twentieth century and a slight increase was registered after 1950 (Williams, 1989; World Resources Institute, 1987).

In most of the DMEs, the growing stock and annual net growth of timber have increased even more dramatically. In the United States, annual net growth more than trebled between 1920 and 1980 (Clawson, 1979), despite predictions in the early 1900s of an impending timber famine. In Europe the forest area showed a net increase of 2 million hectares between 1980 and 1990, whereas the growing stock and net annual increment expanded by 16% and 19% respectively (UNECE/ FAO, 1992; Kuusela, 1994). In short, the forest is an expanding resource, and this distinguishes it in a world supposedly facing an increasing scarcity of natural resources.

The background to forest expansion

Why is the forest resource in the DMEs expanding in both area and volume? The answer probably lies in a combination of circumstances rather than in a single factor. Indeed, multiple or complex causes are suggested by the fact that areal expansion results from both afforestation and from natural regeneration, especially on land abandoned by agriculture. Statistical data on the relative extents are incomplete. Nevertheless, it is clear that in some countries, including the United Kingdom and Ireland, the expansion has been wholly or largely by afforestation (i.e. by active reforestation), whereas in others, particularly Spain, natural regeneration (passive reforestation) has played a very important part. In Europe as a whole, natural regeneration occurred on a gross area of just over 1 million hectares between 1980 and 1990, compared with just under 2 million hectares of afforestation (UNECE/FAO, 1992).

A combination of socio-economic, political and cultural factors underlies the forest transition. One significant factor, at least in some countries such as France and Denmark, was an important shift in the prevailing perception of the forest. Traditionally it had been used by local people for multiple products, including fodder and grazing as well as fuelwood and constructional timber. By the end of the eighteenth century, however, this preindustrial forest paradigm was giving way to one in which the forest was perceived primarily as a source of industrial wood, and its management was increasingly geared to urban-industrial demands. In short, agricultural space and forest space became increasingly separated in both spatial and functional terms.

During the eighteenth and nineteenth centuries, rural population growth had pushed the frontier of settlement and cultivation far into the mountain valleys, bringing new pressures into areas which had previously survived the deforestation that affected the lowlands. During the twentieth century, the process has been reversed with agricultural depopulation, the abandonment of both arable and grazing land, and subsequent forest regeneration. The return of the forest may signal the withdrawal of human pressures, but the resulting landscape changes have not always been welcomed. In addition to this passive reforestation, deliberate reforestation took place in parts of the Alps in the second half of the nineteenth century in the hope of reducing floods (Fairbairn, 1996). Furthermore, timber shortages from the late eighteenth century stimulated the adoption of more effective forest management, in particular the advent of sustained-yield management. Many earlier episodes of timber shortage had been encountered, e.g. in England and in France, but a successful long-term response was first forthcoming in the forests of Prussia. During the nineteenth century, the evolving power of the state was better able than at any previous time to enact and enforce forest laws. State intervention in forestry became common, in tacit recognition that market forces and forest conservation could not comfortably coexist.

One of the underlying influences on state involvement, at least in some countries, was a desire to make productive use of otherwise unproductive land. In Denmark, state encouragement for the afforestation of the barren Jutland heathlands began as early as the mid-nineteenth century (Jensen, 1993). *Production* was usually interpreted in terms of timber destined for industrial use. Earlier perceptions of the forest as the source of a multiplicity of goods and services were by now narrowing; the forest began to be viewed exclusively as a source of timber. Several barren coastlands, notably on the Baltic and in Les Landes in south-west France, were already being extensively afforested by the early nineteenth century (e.g. David, 1994). By the beginning of the twentieth century, an enduring association between afforestation and poor, marginal land had been formed. As late as the 1980s, better qualities of agricultural land were still being protected against afforestation in countries such as the United Kingdom and Germany. As will be seen later in the chapter, this association has not been easily broken in the changed circumstances of the late twentieth century.

State-assisted forest expansion in the twentieth century

If the foundations for modern forest expansion were laid in the nineteenth century, the twentieth century was to witness direct or indirect state efforts at afforestation in numerous developed countries ranging from Ireland to New Zealand. In many countries, the state forest service has directly afforested land and has indirectly encouraged private sector planting through incentives such as grants and tax allowances. Timber shortages, actual or predicted, often underlay this involvement of state enterprise in forest expansion. In Britain the exigencies of World War I were the trigger for radical action in establishing a state forestry service, the Forestry Commission, empowered to acquire and afforest land on behalf of the nation. This trigger operated, however, in the context of a forestry infrastructure (including forestry schools) that had begun to develop in the late nineteenth century. At the other side of the world, New Zealand also began to fear a timber famine, and in the 1920s embarked on a programme of state-assisted afforestation (e.g. Roche, 1990; Roche and Le Heron, 1993).

This state involvement in afforestation has been maintained in many DMEs almost throughout the twentieth century, even if planting rates have fluctuated and the ostensible reasons for continued expansion have varied. A feature of many national afforestation histories has been the volatility of planting rates. Although programmes of expansion have in some cases been sustained over several decades, the rates of expansion have varied, with boom periods alternating with slower phases. During boom periods, underlying tensions between forestry and agriculture and forestry and environmental issues have given rise to controversy and conflict (e.g. Le Heron and Roche, 1985; Mather, 1991a).

In general terms, the new forests were designed for maximum timber production and were located in otherwise unused areas or in areas of land of poor agricultural quality. To this extent, forestry was a residual land user. The primacy of agriculture prevailed, especially in the climate of food shortages in postwar Europe, and the location of new forests was often determined by a desire to minimize their impact on agricultural production. Furthermore, the new forests were generally located in areas remote from the main centres of population.

Forest expansion at the end of the twentieth century

By the latter part of the twentieth century, agricultural surpluses existed in the European Union and the United States, and the protectionist policies pursued by some countries in respect of farmland had weakened. It might be expected, therefore, that afforestation rates would accelerate. Indeed, numerous policy initiatives have been taken since the mid-1980s (earlier in some countries) to encourage the transfer of farmland into forest, and in particular to encourage farm forestry. For several reasons, however, this expectation of accelerating expansion is not fully matched by achievement.

Finland was one of the first countries to implement a policy of farm afforestation in modern times. By the late 1960s it was becoming apparent that Finland's policy of agricultural self-sufficiency was becoming overly successful in that more food was being produced than was required. An area of some 400 000 hectares of farmland was deemed surplus to requirements, and a 'field' afforestation policy was implemented in 1969 in an effort to encourage an outflow of land from agriculture (Selby, 1974; Selby and Petäjistö, 1994). More recently, various programmes have been devised both in European countries and in the United States to encourage the afforestation of farmland. In the United States, farmers are encouraged by financial incentives to afforest erodible land withdrawn from agricultural production, under the Conservation Reserve Program. In Britain, a Farm Woodland Scheme (FWS) was introduced in 1988 and a revised Farm Woodland Premium Scheme (FWPS) in 1992. A powerful obstacle to farmers' involvement in forestry is the long interval between initial investment and eventual return. The FWS and similar schemes operating in other countries therefore offered annual payments for up to 20 or more years, as well as the initial planting grants generally on offer. Under FWS, annual payments of up to £190 per hectare could be paid to farmers, for 40 years for oak and beech, 30 years for other broadleaves, 20 years for other woodland and 10 years for traditional coppice. Similar initiatives emerged in other EU countries. In France, a scheme introduced in 1991 offered annual payments of FF 1000 per hectare for 5 years for short-rotation coppices, 10 years for conifers and fast-growing broadleaves, or 15 years for other broadleaves such as oak or beech (Pelissie, 1992).

By the early 1990s, the drive to reduce the problem of agricultural surpluses by transferring agricultural land into forest was intensifying. In Finland, an area of around 800 000 hectares, or 40% of the 'field' area and 30% of the cultivated area, was now thought to be surplus to requirements in relation to a policy of 100% agricultural self-sufficiency. This was twice the estimate of 1969, when the field afforestation policy was launched. Despite the evident failure of the initial programme to make a significant impact on the area perceived to be surplus, a strengthened 'field' afforestation policy was now launched, with increased payments (Selby, 1994; Selby and Petäjistö, 1994). In the European Union, the CAP reform package agreed in 1992 included a significant 'accompanying measure' in the shape of Regulation 2080/92 (Chapter 4). This regulation strengthened the emerging role of the EU in forestry matters, and included several provisions:

1. Aid for afforestation of agricultural land, to be granted irrespective of land-owner, and intended to cover costs (up to 4000 ecu per hectare for hardwoods, 3000 ecu per hectare for conifers and 2000 ecu per hectare for eucalyptus).
2. Aid for the management of plantations, for five years.
3. Annual premia to 'compensate' for loss of income (for a maximum of 20 years) of up to 600 ecu per hectare for agricultural holdings and 150 ecu per hectare for other landowners.
4. Aid for woodland improvements on agricultural holdings.

While these measures are clearly geared to agricultural policy and to the problem of surpluses, the EU had also become involved in encouraging afforestation within regional development schemes. Perhaps the most notable instance has been in Ireland, where after a slow expansion of forests on agricultural land during the early 1980s, planting rates accelerated dramatically during the second half of the decade, culminating in the Forestry Operation Programme of 1989–1993 (Gillmor, 1992; 1993).

In terms of planting rates, the farm afforestation programme in Ireland has been spectacularly successful. The Irish experience, however, is rather unusual, both in terms of the rapid incremental development of afforestation incentives during the 1980s and more especially in terms of uptake of the incentives and of the area planted. In most other countries, the success of farm afforestation programmes has been more muted, both in terms of area and on effects on agricultural production and surpluses. In England the pattern of adoption of farm woodlands (through FWS) was found by Ilbery (1992) to resemble that of other innovations: larger, younger and better-educated farmers with higher incomes were over-represented among the adopters. Ilbery and Kidd (1992) concluded that many of the farm woodlands might have been planted anyway, and that in general terms the farmers in most need of financial support were not entering the scheme. They also concluded that FWS had made little impact in the main arable areas of eastern England: the main areas of uptake were in more central counties, especially where there was some tradition of farm woodlands.

Ilbery and Kidd went on to conclude that FWS had had little success in achieving three of its aims: to divert land from agricultural production and thus help to reduce agricultural surpluses; to contribute to farm incomes and to rural employment; and to contribute to the United Kingdom's timber requirements by encouraging a greater interest in timber production on farms. Rather more success was achieved in respect of the fourth aim, to enhance landscape. It is perhaps significant that when the Farm Woodland Premium Scheme was introduced in 1992, its primary objective was to 'encourage planting of woodland by farmers, *thereby enhancing*

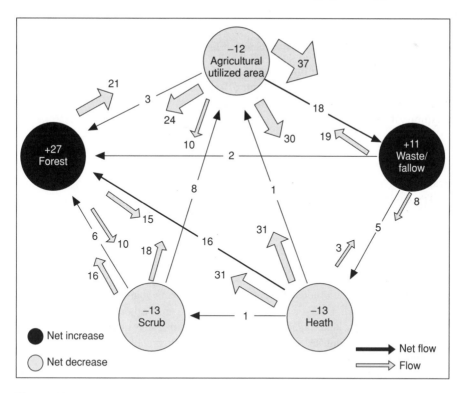

Figure 6.1 Transfers between five categories of rural land in France, 1982–1990
(Reprinted from Cavailhes and Normandin, 1993, by permission of Ecole Nationale du Genie
Rural, des Eaux et des Forêts.)

the landscape' (Ministerial Statement, 31 January 1992, italics added). Just under
14 000 hectares were planted under FWS during the three years of its operation,
compared with a target of 36 000 hectares. In the light of this experience, no target
or quota was set for FWPS.

In general terms, the experience in other countries has been similar. In France,
little impact of farm afforestation schemes was felt in the main arable areas in the
Paris Basin, where indeed the area of farm woodland continued to decline during
the 1980s (Cavailhes and Normandin, 1993). As Fig. 6.1 indicates, the detailed
pattern of land transfer between different categories is complex. Nevertheless, it is
clear that afforestation in France during the 1980s was mostly on unused land or
on relatively unproductive marginal land or heathland, rather than on productive
arable land. Most of the planting was carried out in more marginal agricultural
areas in the south of the country. Although it is true that most of this planting
predated the introduction of explicit farm afforestation policies, it is probable that
similar patterns will be produced by more recent plantings. In Italy, most of the
interest in farm afforestation has been in upland areas, especially in the southern
and central part of the country, rather than in the north. In Germany, interest is
largely restricted to poorer land.

In Finland, the pattern of 'field' afforestation at the local level has been associated with stoniness, waterlogging, low soil fertility and outlying locations. At the regional level, an association has been identified with areas having high degrees of rurality and with areas suffering from poor social conditions and economic backwash effects (Selby, 1980). In general, farm afforestation has had little impact on the most productive arable areas, which contribute most to agricultural surpluses. Furthermore, it is not always the most productive farmers who are likely to be attracted to a farm woodland scheme. The financial incentives for farm afforestation tend to be relatively less attractive on prime land; besides, the ethos which for more than two centuries has associated afforestation with marginal land or wasteland will not evaporate overnight.

The design of effective policy measures to encourage farm afforestation is problematic. The basic premise has attracted surprisingly little critical review: it is that the problem of surpluses should be tackled by reducing the agricultural area, rather than the other factors of production. But even if this premise is accepted, problems still remain. One is the setting of payment levels that are meaningful incentives on the more productive areas without being excessively attractive on more marginal land. Although a crude differentiation of payments is attempted in the United Kingdom – between Less Favoured Areas (LFAs) and other land – the precise gearing of levels to land capability would be administratively very difficult. Flat-rate payments in countries such as France and Italy are likely to have spatially differentiated effects.

It seems that policy makers have assumed the key to the success or otherwise of a farm afforestation scheme is its level of payments. The financial attractiveness of a scheme is undoubtedly important; for example, the inability of the annual payments offered under FWS to compete with alternative sources of revenue was identified by Edwards and Guyer (1993) as a constraint in Northern Ireland. But it is by no means the only significant variable. Social and psychological factors may also be significant. In central Scotland, Clark and Johnson (1993) found that social factors, such as the increased public access and vandalism which farm woodlands might attract, were perceived by some farmers as obstacles to planting. They also found that farmers generally perceived forestry to be an enterprise quite separate from farming, and that two-thirds of them were simply opposed to tree planting on productive agriculture land. They had deeply embedded psychological and moral reasons for focusing on food production. Their working lives had been set in a context of the primacy of agricultural production, in which it was assumed that agricultural land should be preserved from encroachment by forests or any other form of land use which did not produce food.

Forestry simply conflicted with the production-oriented ethos of farming. Although a policy switch could be made quickly (as indeed had happened around 1985), the attitudes of the farming population are less malleable and do not change quickly. Twenty years previously, Selby (1974) had highlighted the difficulties of Finnish farmers in accepting the rapid changes in agricultural policies and objectives associated with the introduction of the first field afforestation programme in 1969: 'old values have been destroyed and new ones have yet to be accepted' (p. 6).

More recently, non-economic factors such as a desire to retain the family farm as a family farm and a sense of attachment to place have also been identified as significant constraints on farm afforestation in Finland (Selby and Petäjistö, 1994). The provision of technical advice and support to farmers embarking on the new venture of tree growing may also have been perceived as inadequate, e.g. in Italy (Nunzi, 1992). In short, a range of non-economic factors may be significant in relation to the uptake of incentives for farm afforestation. Perhaps these factors have not always been adequately considered in the design of farm forestry schemes.

Paradoxically, at a time when determined efforts have been made to transfer land from agriculture to forestry, afforestation rates in countries such as the United Kingdom, Italy and France have declined. Furthermore, the area of farm woodlands in France has actually been decreasing, and the rate of decrease has been greater for farm woodlands than for the total agricultural area. Between 1979 and 1988, the area of farm woodlands decreased by 17% and the total agricultural area decreased by 3% (Cinotti, 1992). Farmers have been abandoning their woodland, in the context of an increasing degree of agricultural specialization. As a result, the traditions of farm forestry have been weakening and forestry skills have been lost (Pelissie, 1992).

Farm afforestation policies have therefore been introduced against a background in which the long-term trend is unfavourable, despite the apparently favourable circumstances for the transfer of agricultural land into forest. The intensification of agricultural production in the main arable areas has been accompanied by reductions in the area of farm woodlands, and in the short term at least there is little prospect this trend will be reversed. In countries where there is a strong tradition of integration of farming and forestry, agricultural change has meant the links have tended to weaken. This trend has been associated in particular with changing patterns of ownership, as ownership by farmers declines relative to ownership by other non-industrial private owners. In other countries where there is little or no tradition of farm forestry, attitudes are at best slow to change, and involvement in farm woodland schemes is more likely to be manifested in small-scale afforestation of areas of marginal land than in meaningful integration.

As yet, farm afforestation has had little impact on agricultural surpluses, nor has it made much impression on formally 'set aside' agricultural land. In Scotland, only around 5% of the 'set aside' area was under woodland by 1992, and the corresponding percentage on Italian set-aside land was under 4% (Nunzi, 1992). Overall planting rates, both within farm woodland schemes and elsewhere, have declined in several countries in recent years, despite apparently favourable circumstances for forest expansion. In France, rates of state-assisted afforestation declined from an annual average of 30 000 hectares in the 1960s to under 10 000 hectares in the 1980s (Cavailhes and Normandin, 1993). In the United Kingdom they declined sharply after 1988, when tax incentives for afforestation were suddenly withdrawn (Fig. 6.2). One underlying factor, especially in the United Kingdom, is changing forest values and perceptions. Whereas a hundred years ago afforestation was generally regarded as environmentally beneficial, it is now seen as detrimental, at least if carried out in certain settings and in certain ways.

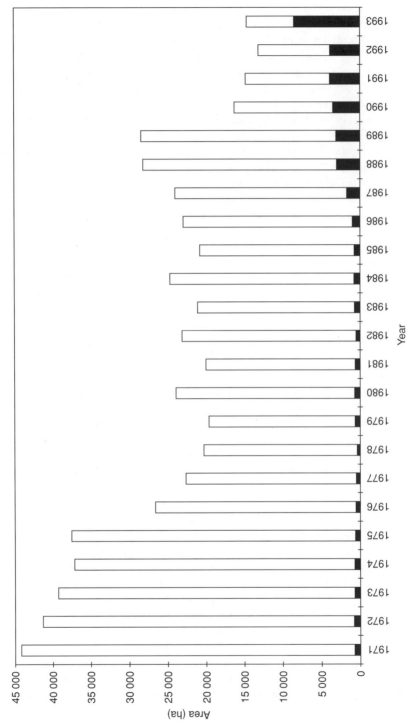

Figure 6.2 New planting in Great Britain, 1971–1993: (□) conifer and (■) broadleaf (Compiled from annual reports of the Forestry Commission)

CHANGING FOREST VALUES AND PERCEPTIONS

As material needs in the developed world have been satisfied in terms of food, shelter and consumer goods, people have looked to the forest for more than wood. Demands have grown for services such as recreation and nature conservation, and non-material forest values have become more prominent. The symbolic value of the forest is great. At one level, the forest symbolizes the environment: a clear-felled forest or one suffering from decline from disease or air pollution is symbolic, in the eyes of many, of an environmental malaise or a dysfunctional human–environment relationship. At another level, it symbolizes a national heritage: Kennedy (1985) refers to the nostalgic and nationalistic imagery of pristine forests in countries such as Sweden, the United States and Germany; in relation to Germany, see also Schama (1995). In the same way, the oakwood of England or the native pine forest of Scotland have a symbolic value, irrespective of their value as timber. It is not surprising, therefore, that forests have become battlegrounds for value systems. This is especially so in an era of rapid change in rural society as well as in society as a whole, and in an era of unprecedented concern about the state of the environment.

Recent decades have witnessed a fundamental shift in the nature of forest values and perceptions held by society in many DMEs. In short, the value of the forest as an environment has increased relative to its value as a source of timber; there has been a swing from instrumental value to intrinsic value. This shift has been most noticeable in the 'old growth' native forests of the new lands, both in North America and Australasia, but has not been restricted to such areas. It has also occurred in European forests, in countries such as Germany and the United Kingdom (e.g. Mather, 1991b). It has occurred in the context of plantations in countries such as the United Kingdom, even if it has not yet greatly affected plantations in the new lands (in general 'tree farms' are still acceptable in the new lands but not in the United Kingdom and similar countries). It encompasses both the functions or purposes of forests, and the approach adopted to their management. In essence, a paradigm shift has been occurring. Its implications are profound and extend to national and international forest policies as well as to the types and locations of new forests being created by afforestation.

Public values in the United States, and subsequently in numerous other countries, began to change after World War II, especially during the 1960s (e.g. Hays, 1987). Memories of scarcities and hardships of the war years and of the prewar Depression receded; material abundance was taken for granted, but now seemed insufficient. Furthermore, Western society had become increasingly urban, and as it did so, its environmental attitudes and value systems changed. In the words of Kennedy (1985), 'Consumptive values are commonly associated with agricultural/rural society and non-consumptive values with urban society' (p. 124). Consumptive values (especially those associated with timber production) were now strongly challenged by intangible non-consumptive or 'environmental' values. Since the nineteenth century, timber production had been seen as the main objective of forest management throughout much of the developed world. Although protection forests were extensive in some areas, the main forest value was usually perceived to be economic, in the form of the production of a raw material for industry. Other values, perhaps for wildlife or recreation, were usually seen as secondary or subsidiary, if they were recognized at all. Forest management was essentially a technical exercise, geared to clear goals and based on economic values.

By the 1960s, the challenge to this paradigm was mounting. In general, changing forest values reflected the questioning of the industrial growth paradigm and other societal changes that characterized the 1960s. In particular, intensive forestry practices could now alter landscapes more radically, and at the same time a growing number of environmentalists and recreationists could witness and criticize the impacts of these practices (Reunala, 1984). One of the first signs of a challenge to the established forest management paradigm was the growing demand for outdoor recreation in the federal forests of the United States. In 1960 the Multiple Use-Sustained Yield Act was passed. Some have concluded that it paid only lip service to the concept of multiple use, and believe its practical significance was limited. Nevertheless, it stands as a milestone on the road to a new management paradigm.

Numerous other indicators of progress along the same road can be quoted. For example, in New Zealand state forests, soil and water conservation were added to the original function of timber production during the mid-1950s, recreation in the 1960s, nature conservation in the early 1970s, landscape conservation in the late 1970s and the provision of educational opportunities in the early 1980s (Tilling, 1988). In the Netherlands, recreation was added to the management objectives of state forests in the 1960s and nature conservation in the 1970s (Grandjean, 1987). Legislation included the Native Forest Act in New Zealand in 1976, the Forest Resource Management Act in the United States in the same year, and the Preservation of Beech Forests and Forestry Act in Sweden in 1979. Between the 1970s and 1990s, the manifestations of changing societal values were becoming unmistakable; societies in many DMEs were now holding perceptions and expressing demands which did not accord easily with the traditional management paradigm and the assumption of timber primacy.

One of the battlegrounds in the clash between the new values and the old was in the Pacific North-West of the United States. The conflict was between logging interests and environmentalists. The northern spotted owl was the focus of the conflict, but in a wider sense, it became a metaphor for the new values. A bitter struggle eventually resulted in effective victory for the environmentalists over the loggers (e.g. Watson and Muraoka, 1992). Large areas of federal old-growth forest were withdrawn from logging, and logging in the national forests of the Pacific North-West fell from 5.2 billion board feet in 1989 to 2.1 billion in 1992 (Hirt, 1994). Similar struggles over old-growth native forests occurred in British Columbia, Australia and New Zealand. Even if the specific focus varied from place to place and the immediate outcome was not always the same, the trend was obvious; native forests were now being valued for reasons other than timber.

Although the shift in values and the paradigmatic shift in management are perhaps most clearly demonstrated by the native forests of the new lands, they are also in evidence in the planted forests of the old world. In the United Kingdom, a similar trend was clearly evident during the 1980s. From the outset of state-aided afforestation at the end of World War I, the primary objective was to grow timber. To this end, conifers such as Sitka spruce were the preferred species. Most of the planting was on previously bare heathland or moorland, but some areas of broad-leaved woodlands were replanted with conifers. And some areas of broadleaved woodlands were by the 1960s and 1970s also suffering agricultural encroachment. The result was that by the late 1970s and early 1980s, the surviving area of broad-leaved woodland was dwindling. In response to pressure from environmental inter-ests, a 'broadleaved' policy was introduced, embracing a presumption against further 'coniferization' of broadleaved woodlands and a new grant scheme to encourage the planting of broadleaved species. All previous planting-grant schemes required that timber production should be the primary objective: the new Broadleaved Wood-land Grant Scheme did not have this requirement.

By the mid-1980s, various aspects of 'industrial' afforestation, including the means of ground preparation and the design and species composition of the new plantations, were attracting criticism from state agencies such as the Nature Con-servancy Council and the Countryside Commission for Scotland, as well as from interest groups such as the Royal Society for the Protection of Birds. In 1988 a presumption against commercial coniferous planting in the English uplands was announced. Furthermore, controversy over the afforestation of bogland in northern Scotland eventually led in 1988 to the radical removal of the tax concessions that had been the main driving force for private sector afforestation. Neither the ends (in the sense of timber primacy) nor the means (in the sense of the nature of the incentives and the techniques of establishment) of long-established policy were now acceptable. In the case of planted forests in the United Kingdom, as in the old-growth forests of the Pacific North-West, environmental values were overcoming those of timber production.

Numerous terms have been applied to the emerging paradigm that has chal-lenged traditional forestry management. They include new forestry, new perspect-ives, holistic forestry, kinder and gentler forestry, sustainable forestry, multi-value forest management, multi-resource forest management, and forest ecosystem man-agement (Bengtson, 1994). The various terms, and indeed the general paradigm (or paradigms) to which they are related, are difficult to define precisely, but they are characterized by several features which distinguish them from the more traditional paradigm. Some contrasting features are summarized in Table 6.2.

The new paradigm does not mean that old values and objectives have been completely abandoned, but simply that new values and objectives have grown in relative strength. The extent of the shift varies between countries and between different forest areas within individual countries. That a shift has been in progress in the second half of the twentieth century cannot be doubted. It is symbolized in various ways. For example, the United States Forest Service formally espoused the 'forest ecosystem management' paradigm in 1992 (Brown and Harris, 1992). Some years earlier, the passing of the Forest Act of 1975 in Austria marked the end of an era by terminating the principle of timber primacy enshrined in the Forest Act of 1852 (Glück, 1987). Just as the earlier act was representative of its period in Europe

Table 6.2 Management paradigms: traditional and 'forest ecosystem'

	Traditional	Forest ecosystem
Philosophical base	Utilitarian	Environmental ethic
Values	Instrumental	Intrinsic
Objectives	Timber production	Multiple: goods and services
Role of science	Management as applied science	Management as applied natural and social science, informed by value system
Themes	Mechanistic, reductionist view Timber famine	Systems, holistic view Biodiversity loss

Source: Adapted from Bengtson (1994), Behan (1990) and Brown and Harris (1992).

Table 6.3 Typical characteristics of industrial and post-industrial forests

	Industrial	Post-industrial
Size	Large	Small
Species composition	Monocultural Conifers	Diverse Broadleaves
Location	Remote, peripheral	Lowland, peri-urban
Management objectives	'Material' Timber production	'Environmental' Wildlife conservation, recreation

and marked the transition from folk use of the forest for a variety of purposes to intensive management for timber production, so now a shift back to diversity of function was taking place.

The term *post-industrial forest* has been used by Mather (1990) to characterize the physical manifestation of the changing management paradigm. Although the primacy of timber production is the characteristic of the industrial forest, in the post-industrial age there is also a demand for other forest goods and services. In particular, these services include recreation and wildlife conservation. Some comparative characteristics of industrial and post-industrial forests are summarized in Table 6.3. As Table 6.3 suggests, both the composition and the location of 'post-industrial' plantations may differ from those of the more traditional 'industrial' forest. The increasing relative importance of broadleaved species in British afforestation is clearly illustrated in Fig. 6.2. From negligible areas at the beginning of the 1980s, the use of such species rose to around half of the total new-planting area by the 1990s.

In the traditional management paradigm, the conventional wisdom was that slow-growing broadleaved species were uneconomic; now sufficient inducements are offered in the form of planting (and management) grants to encourage their use. A further contrast lies in location. From the beginning of the British afforestation programme around 1920, the afforestation was largely restricted to non-arable land, and increasingly from the 1960s was concentrated in the remoter rural areas. By the late 1980s, two major changes had combined to begin to reverse that pattern. One is the removal of policies of preservation of arable land (in the context of agricultural surpluses); the other is the increasing relative importance of the recreational function of forests. By the early 1990s, higher planting grants were being offered for planting on arable land, which by its very nature tends to be located in the lowlands, and also for planting in the vicinity of urban areas.

One of the most striking manifestations of a changing paradigm is the concept of the national and community forests. The idea of a national forest in England was promoted by the Countryside Commission from 1987. The potential for multi-purpose forestry in the lowlands lay at the heart of the idea, and the national forest was envisaged as an extensive area of mixed woodland interspersed with farmland. Mixed-species timber would be produced, a recreational resource would be created and environmental enhancement would occur. An area in the Midlands amounting to 50 000 hectares was chosen in 1990 as the site for the proposed national forest. Most of the land is at present farmed and in private ownership, and the existing area of woodland is only 3000 hectares (Bell, 1993). With extensive public acquisition of the land being politically unrealistic at present, the success will depend on the effectiveness of incentives in the form of planting and other grants, and on the success of the lead body, the Countryside Commission, in stimulating interest. A related proposal, originated at the same time, was for a series of community forests to be established near major cities. Their primary purpose would be environmental enhancement. Partnerships between the Countryside Commission and local authorities have been envisaged as the means of establishing and managing the community forests. It is significant that the promoter of the national and community forests is not the state forestry service (the Forestry Commission) but the Countryside Commission, which is the state body (in England) with responsibility for countryside recreation and landscape conservation.

Whatever level of success may be achieved in implementing the concepts of the national and community forests, there is little doubt that the preferred composition and location of forests in England have undergone radical change. The trends reflected here, however, are not unique to that country. In the Netherlands, the Forestry Plan of 1984 aimed at expanding the forest area by around 10% by the end of the century, and one-third of the expansion was to be near the main cities (van den Berg, 1989). In small, densely populated countries non-material forest functions are already of fundamental importance. Quantification of the relative importance of different forest functions is fraught with difficulty, because of the lack of adequate yardsticks, and the results of measurement efforts need to be interpreted with caution. Nevertheless, recreation is deemed to be of 'high' importance in 80% of the state forest area in the Netherlands, and on 50% of the private forest. In comparison, timber production is of 'high' importance on 30% of both state and private forest (UNECE/FAO, 1992). In contrast, timber production is of 'high' importance in 87% and 98% respectively of public and private forests in Finland, and recreation is of 'high' importance in none of the area.

This comparison, however, does not mean that the Finnish forest is valued only for its timber nor that heavily forested countries have escaped the trends that have been operating in countries such as England and the Netherlands. In Finland, as indeed in other countries with traditions of extensive farm-forest ownership, rural changes have meant that numbers of 'other' non-industrial private owners have increased relative to numbers of farmer-owners. The decline of the agricultural population and a drift to the cities have meant that many former farm forests have passed into the hands of non-farmers. This has often been by inheritance, as a property has passed to the heirs of a farm-forester. Although the farm-forester may have perceived the forest primarily as a source of income, the successors may be more interested in a pleasant environment (for a second home or for a residence combined with non-agricultural employment). The economic significance of the forest has thus altered.

This trend has been in operation for several decades in countries such as Finland. For example, in the central Finnish province of Keski-Suomi, forest owner-ship by non-farmers increased from 8% of the area in 1945 to 20% in 1970 (Reunela, 1974). With increasing numbers of these holdings being used for recreation or residence, the result is likely to be a reduction in timber production. Related social trends may have a similar effect; timber harvesting tends to decrease with increas-ing age of owner (Kuuluvainen and Salo, 1991), and an ageing farming population might in itself mean a declining timber supply even in the absence of ownership changes. Furthermore, the state of agriculture can have significant implications for the level of timber production. In Austria, economic imperatives have driven many small farmers to seek off-farm employment, with the result that many farm forests are neglected (Eckmüllner, 1986).

Similar trends are also apparent elsewhere. In Pennsylvania, only 1% of (NIPF) owners held their land primarily for timber production, and the correponding figure for the neighbouring state of Maryland was only 2% (Birch and Dennis, 1980; Kingsley and Birch, 1980). Most owners sought a pleasant setting for their resid-ence or recreation, rather than direct economic benefits. In New Hampshire, Dennis (1989) found that timber harvesting was negatively correlated with years of formal education; better educated owners perhaps value forest amenities more highly and they may have higher incomes, making income from timber sales less important. Overall in the United States, it has been estimated that only around 5% of small owners manage their forest for timber production, using all or most practicable forestry practices (Cunningham, 1982). In short, timber removals in countries with significant levels of small-scale ownership are often much lower than biological production would permit. In this sense, they may be sub-optimal. Suboptimality stems from a variety of reasons operating individually and in combination. These include fragmented ownership with its diseconomies of scale, changing patterns of ownership and more general rural social change. Changes in forest values and perceptions which highlight the value of the forest as an environment, rather than as a timber resource, are superimposed on these other factors, and compound them. And since the forests are not the main source of income for many NIPFs, their management is not sensitive to grant aid and policy instruments of similar kind.

AGRICULTURE AND FORESTRY

A comparison of the agriculture and forestry sectors in the DMEs at the end of the twentieth century reveals both similarities and differences. Both can be said to have entered an era of 'post-productivism', not in the sense that production is no longer sought but rather in the sense that it is no longer paramount. Environmental values have strengthened in both sectors. Both activities have a bearing on the 'countryside' that is so valued by urban society. That value may be both practical, in terms of amenity and pleasant residential or recreational environments, and symbolic, as the perception by the naive or nostalgic of a mythical golden age with a simpler and better lifestyle.

The transition to an era of post-productivism has not been smooth and uneventful in either sector (Chapter 4). Farmers and forest owners, and the agricultural ministries and forest services that support them, have not found it easy to jettison the ethos and mindset of striving for production goals and the belief in the ability of science and technology to solve management problems. In the case of agriculture, the economic imperatives of support costs have been a major driving force for the transition, whereas environmental values have played a similar role in relation to forestry. There are also other contrasts. In particular, overproduction and surpluses are perceived as problems in agriculture but not in forestry. Indeed, one well-established problem in forestry is to persuade NIPF owners to harvest more timber. On the other hand, both sectors have undergone extensive mechanization since World War II and both have suffered from massive declines in employment. Their economic role in the countryside has therefore decreased markedly, just as their environmental role in contributing to the character of the 'countryside' has increased.

PROSPECT

National forest policies have been in a state of flux in the latter part of the twentieth century. On the one hand, there have been some signs of closer integration between agricultural and forest policies. On the other hand, they have incorporated some of the values associated with the post-industrial forest, with forest ecosystem management and with sustainability. The transition has not been easy, and the period of adjustment is not yet over. In Britain, substantial changes in the administration, implementation and financing of policy measures will be required if the consensus policy objectives are to be achieved (Kanowski and Potter, 1993). One particular issue is the extent to which national policies can be applied uniformly throughout a country. In the United Kingdom, a parliamentary committee has urged that separate policies should be devised for England, Scotland, Wales and Northern Ireland, in order to reflect the different conditions and cirumstances in these countries (House of Commons Environment Committee, 1993). At the same time as spatial refinement of policies is being sought, however, international and global influences are being strengthened. The traditional national forest policies that have prevailed for most of the twentieth century have therefore come under pressure from two directions.

In the nineteenth century, near-revolutionary change occurred in the perception and management of forests in much of Europe. In essence, the spectre of timber shortage had stimulated the adoption of intensive management and the assumption of timber primacy. Forests and their conservation as sustainable sources of timber were key themes in the American Conservation Movement, which reached its zenith during the first decade of the twentieth century. Perhaps history will show the late twentieth century to be a period of comparable significance, in which new perceptions, new objectives and new assumptions emerged. National forest laws were enacted in many countries in the mid-nineteenth century. It would probably be a futile exercise to attempt to evaluate their relative contribution in the transition towards expanding forests and a growing forest resource during the twentieth century, but it can reasonably be concluded that they were one significant factor. Perhaps the late twentieth century will be associated with international forest law in the same way in which the nineteenth century was associated with national law, and perhaps it may also be associated with a transition towards sustainable 'forest ecosystem management' in the same way as the earlier period is associated with the beginnings of sustained-yield management. The past few years have seen much discussion of the tenets of sustainable forest managment, and indeed 1994 saw the publication of a journal article entitled 'International law of forests' (Hooker, 1994).

One of the Forestry Principles emanating from the United Nations Conference on Environment and Development at Rio in 1992 was that forest resources should be sustainably managed to meet the social, economic, ecological, cultural and spiritual needs of present and future generations. In subscribing to these principles, national governments acknowledge that forests have values ranging beyond the economic values of timber production and endorse the concept of sustainable management. To be meaningful, these concepts have to be put into practice, and it is too early to say whether the late twentieth century will witness a major turning-point in forest policy and management. Nevertheless, there are some signs of progress, not only in terms of restatements of national forest policies but also on the international scene.

Since 1990 the Helsinki Process has begun to develop general guidelines for the sustainable management of forests in Europe and to devise measurable criteria and indicators that can be employed in relation to that management. The parallel Montreal Process, involving a number of countries fringing the Pacific, began in 1993 and resulted in the Santiago Declaration of 1995. One of its criteria is the 'maintenance and enhancement of long-term multiple socio-economic benefits to meet the needs of societies'. Among the indicators suggested in relation to that criterion are measures of recreation and tourism, of employment and community needs, and of cultural, social and spiritual needs and values. Much remains to be done in devising operational indicators, as well as in applying the basic principles. Nevertheless, the degree of activity at the international level has been impressive in recent years.

This is especially true in the European Union. Forestry matters were not included in the Treaty of Rome and there has been no common forestry policy to parallel that of agriculture and fisheries. But in 1992 the European Parliament took the initiative in beginning work on what may become the foundation of a common policy and which may provide the framework within which the various community measures related to forestry may eventually be integrated. One of the key features is that the policy is 'global'; it is recognized that a policy cannot be meaningfully implemented internally within the EU unless it embraces issues such as trade and the role of the EU in relation to tropical forests and forests in other lands (European Parliament, 1994). The EU and many of its member states are major importers of forest products. Perhaps it is salutary to note in conclusion that the changing role of forests in the DMEs has been facilitated by the exploitation of forests elsewhere. Without the availability of imports of forest products, would the growth of environmental forest values in the DMEs have been so pronounced? And would the drive to afforest 'surplus' agricultural land have been stronger?

ACKNOWLEDGEMENT

This chapter is informed by research carried out under the Global Environmental Change programme of the UK Economic and Social Research Council, whose support is gratefully acknowledged.

REFERENCES

Behan, R.W. (1990) Multiresource forest management: a paradigmatic challenge to professional forestry. *Journal of Forestry*, **88**(4), 12–18.

Bell, S. (1993) How fares the national forest? *Quarterly Journal of Forestry*, **87**, 124–28.

Bengston, D.N. (1994) Changing forest values and ecosystem management. *Society and Natural Resources*, **7**, 515–34.

Birch, T.W. and Dennis, D.F. (1980) *The forest-land owners of Pennsylvania*. USDA Forest Service Research Bulletin NE-66.

Brown, G. and Harris, C.C. (1992) The Forest Service: toward the new resource management paradigm? *Society and Natural Resources*, **5**(3), 231–45.

Cavailhes, J. and Normandin, D. (1993) Déprise agricole et boisement: état des lieux, enjeux et perspectives dans cadre de la réforme de la PAC. *Revue Forestière Française*, **45**, 465–81.

Cinotti, B. (992) Les agriculteurs et leurs forêts. *Revue Forestière Française*, **44**, 356–64.

Clark, G.M. and Johnson, J.A. (1993) Farm woodlands in the Central Belt of Scotland: a socio-economic critique. *Scottish Forestry*, **47**, 15–24.

Clawson, M. (1979) Forests in the long sweep of American history, *Science*, **204**, 1168–74.

Cunningham, G.R. (1982) Private non-industrial forests. In Young, R.A. (ed) *Introduction to forest science*. John Wiley, New York, pp. 313–33.

David, R. (1994) La fixation des dunes maritimes de Gascogne. *La Vie des Sciences*, **11**, 123–47.

Dennis, D.F. (1989) An economic analysis of harvest behaviour: integrating forest and ownership characteristics. *Forest Science*, **35**, 1088–1104.

Eckmüllner, O.S. (1986) Die Forstwirtschaft im Grünen Bericht und im Grünen Plan. *Centralblatt für das Gesamte Forstwesen*, **103**, 187–210.

Edwards, C. and Guyer, C. (1992) Farm woodland policy: an assessment of the response to the Farm Woodland Scheme in Northern Ireland. *Journal of Environmental Management*, **34**, 197–209.

European Parliament (1994) *A global Community strategy in the forestry sector.* European Parliament, Luxembourg.

Fairbairn, J. (1996) *The forest transition in France.* Modelling the Forest Transition Working Paper 5, Department of Geography, University of Aberdeen.

Gillmor, D. (1992) The upsurge in private afforestation in the Republic of Ireland. *Irish Geography*, **25**, 89–97.

Gillmor, D. (1993) Afforestation in the Republic of Ireland. In Mather, A. (ed) *Afforestation: policies, planning and progress.* Belhaven, London, pp. 34–48.

Glück, P. (1987) Social values in forestry. *Ambio*, **16**(2/3), 158–60.

Grandjean, A.J. (1987) Forêts domaniales et politique forestière néerlandaise depuis 1954: évolution et résultats. *Revue Forestière Française*, **39**, 219–30.

Hays, S.P. (1987) *Beauty, health and permanence: environmental politics in the United States 1955–1985.* Cambridge University Press, Cambridge.

Hirt, P.W. (1994) *A conspiracy of optimism: management of the national forests since World War Two.* University of Nebraska Press, Lincoln NE.

Hooker, A. (1994) International law of forests. *Natural Resources Journal*, **34**, 823–78.

House of Commons Environment Committee (1993) *Forestry and the environment*, HC 257 1992–93. HMSO, London.

Ilbery, B. (1992) State-assisted farm diversification in the United Kingdom. In Bowler, I.R. Bryant, C.D. and Nellis, M.D. (eds) *Contemporary rural systems in transition*, vol 1, *Agriculture and environment.* CAB International, Wallingford, pp. 100–118.

Ilbery, B. and Kidd, J. (1992) Adoption of the Farm Woodland Scheme in England. *Geography*, **77**, 363–67.

Jensen, K.M. (1993) Afforestation in Denmark. In Mather, A. (ed) *Afforestation: policies, planning and progress.* Belhaven, London, pp. 49–58.

Kanowski, P.J. and Potter, S.M. (1993) Making British forest policy work. *Forestry*, **66**, 233–47.

Kennedy, J.J. (1985) Conceiving forest management as providing for current and future social value. *Forest Ecology and Management*, **13**, 121–34.

Kingsley, N.P. and Birch, T.W. (1980) *The forestland owners of Maryland.* USDA Forest Service Research Bulletin NE-63.

Kuuluvainen, J. and Salo, J. (1991) Timber supply and life cycle harvest of nonindustrial private forest owners: an empirical analysis of the Finnish case. *Forest Science*, **37**, 1011–29.

Kuusela, K. (1994) *Forest resources in Europe*, European Forest Institute Research Report 1. Cambridge University Press, Cambridge.

Kuusela, K. and Salminen, S. (1991) Forest resources of Finland in 1977–1984 and their development in 1952–1980. *Acta Forestalia Fennica*, **220**, 1–84.

Le Heron, R.B. and Roche, M.M. (1985) Expanding exotic forestry and the extension of a competing use for rural land. *Journal of Rural Studies*, **1**, 211–29.

Mather, A.S. (1990) *Global forest resources.* Belhaven Press, London.

Mather, A.S. (1991a) The changing role of planning in rural land use: the example of afforestation in Scotland. *Journal of Rural Studies*, **7**, 299–309.

Mather, A.S. (1991b) Pressures on British forest policy: prologue to the post-industrial forest? *Area*, **23**, 245–53.

Nordic Council (1990) *Yearbook of Nordic Statistics.* Nordic Council, Copenhagen.

Nunzi, L. (1992) Opzione imboschimento: operazione conveniente? *Terra e Sole*, **47**, 604, 639.

Osaka, M.M. (1983) Forest preservation in Tokugawa Japan. In Tucker, R.P. and Richards, J.F. (eds) *Global deforestation and the nineteenth century world economy*. Duke University Press, Durham NC, pp. 129–45.

Pelissie, D. (1992) La prime au boisement des terres agricoles. Quels enjeux pour l'avenir? *Revue Forestière Française*, **44**, 479–89.

Reunala, A. (1974) Structural change of private forest ownership in Finland. *Communicationes Instituti Forestalis Fenniae*, **82**(2), 1–70.

Reunala, A. (1984) Forest as symbolic environment. *Communicationes Instituti Forestalis Fenniae*, **120**, 81–85.

Roche, M.M. (1990) The New Zealand timber economy. *Journal of Historical Geography*, **16**, 295–313.

Roche, M.M. and Le Heron, R.B. (1993) New Zealand: afforestation policy in eras of state regulation and de-regulation. In Mather, A. (ed) *Afforestation: policies, planning, progress*. Belhaven, London, pp. 140–62.

Schama, S. (1995) *Landscape and memory*. HarperCollins, London.

Selby, J.A. (1974) Afforestation of fields in Finland: agricultural background and recent achievements. *Communicationes Instituti Forestalis Fenniae*, **82**(4), 1–51.

Selby, J.A. (1980) Field afforestation in Finland and its regional variations. *Communicationes Instituti Forestalis Fenniae*, **99**, 1–126.

Selby, J.A. (1994) Primary sector policies and rural development in Finland. In Gilg, A.W. (ed) *Progress in rural policy and planning 4*. John Wiley, Chichester, pp. 157–76.

Selby, J.A. and Petäjistö, L. (1994) Field afforestation in Finland in the 1990s. *Finnish Forest Research Institute Research Paper 502*.

Tilling, A.J. (1988) Multiple-use indigenous forestry on the west coast of South Island. *New Zealand Forestry*, 32, 13–18.

Totman, C. (1986) Plantation forestry in early modern Japan: economic aspects of its emergence. *Agricultural History*, **60**, 23–51.

UNECE/FAO (1992) *The forest resources of the temperate zone* (2 vols). United Nations, New York.

van den Berg, L.M. (1989) Rural land-use planning in the Netherlands: integration or segregation of functions? In Cloke, P. (ed) *Rural land-use planning in developed nations*. Unwin Hyman, London, pp. 47–75.

Varjo, U. and Tietze, W. (1987) *Norden: man and environment*. Gebruder Borntraeger, Berlin.

Watson, R.B. and Muraoka, D.D. (1992) The northern spotted owl controversy. *Society and Natural Resources*, **5**, 85–90.

Williams, M. (1989) *Americans and their forests: an historical geography*. Cambridge University Press, Cambridge.

World Resources Institute (1987) *World resources*. Better Books, New York.

PART 3

CHANGING RURAL ECONOMY
AND SOCIETY

CHAPTER 7

RURAL MIGRATION AND DEMOGRAPHIC CHANGE

Gareth Lewis

INTRODUCTION

Migration, or the movement of people from one geographical location to another, has increased in volume and diversity over time and involved steadily lengthening distances. With the emergence of a world economy and the globalization of communications, migration in turn has 'exploded' at all geographical scales and become of major concern, thus justifying Goldstein's (1976, p. 424) observation that, 'whereas the study of fertility dominated demographic research in the past several decades, migration may well have become the most important branch of demography in the last quarter of the twentieth century'. In order to enhance an understanding of these expanding migratory flows, Zelinsky (1971) has linked them to the process of modernization in such a way that different types of migration could be seen as being characteristic of different stages of development. The model highlights two significant transformations in the redistribution of population since 1850: first, during the late nineteenth century the concentration of population, or urbanization, consequent upon the advance of industrialization, into a limited number of areas; and second, during more recent decades, a rather different and what some have regarded as an unexpected trend, whereby 'a number of major centres of population concentration in the industrial nations began to experience a decline in the in-movement of population from the more remote and peripheral regions of those nations. This decline has continued . . . and in many places has gone as far as to create a net flow of population out of major conurbations back into the peripheral and predominantly rural regions' (Vining and Kontuly, 1978, p. 49).

Significantly, the reversal in the fortunes of rural populations has had very different implications to those of urbanization for both the settlement system and the social system. For example, Rowland (1979) and Wardwell (1977) in Australia and the United States, respectively, suggested that migration maintains rather than modifies a settlement system. Essentially, migration is viewed as an equilibrium-seeking process, so that 'migration equilibrium may be approached towards the end of the mobility transition as all parts of the settlement system achieve a high level of modernization' (Rowland, 1979, p. 11). Similarly, in its role as an agent of social change, migration also tends to reduce the difference in the culture, lifestyle and

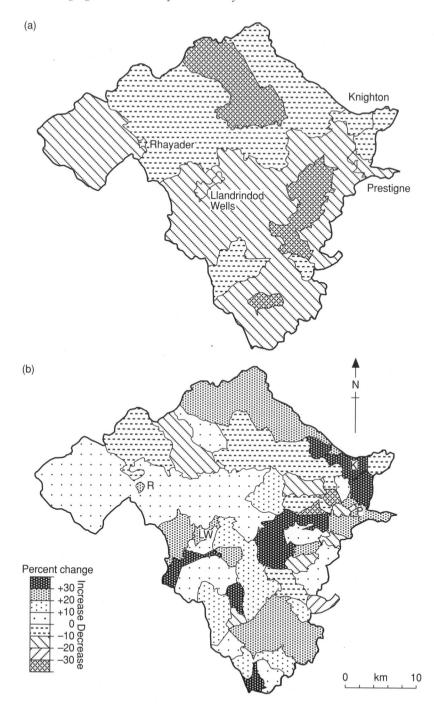

Figure 7.1 Population change in Radnorshire, central Wales:
(a) 1921–1931 and (b) 1971–1981

composition of urban and rural populations (Pahl, 1966; Lewis, 1979). Effectively, contemporary rural migrations reinforce what Parker (1970, p. 4) has described as the postwar 'social avalanche which swept away the last traces of 'village life' and transformed life for everybody, everywhere. . . . There is no point whatsoever in talking any longer about 'village life' and 'farm life'. It's just life.'

This chapter is concerned with recent migration and social change in the countryside and, after a brief consideration of some of the antecedents of contemporary rural migration, it will focus on four interrelated themes:

- The phenomenon of counterurbanization and the rural turnaround.
- Recent and future trends in rural migration and demographic change.
- The significance of the household and household turnover in the migration process.
- A behavioural interpretation of contemporary rural migration and social change.

DEPOPULATION AND SUBURBANIZATION

It is a well-established truism that a better understanding of any transformation, in this case counterurbanization and the rural turnaround, can best be achieved within the broad sweep of past historical events. Lewis and Maund (1976) have suggested a time–space framework, incorporating a changing space-economy and a changing urban–rural relationship, which emphasizes the role of differential migrations over time in effecting demographic change in the countryside. In such terms, the contemporary rural turnaround has its antecedents in two earlier phases, one largely involving rural depopulation and the other involving the onset of widespread suburbanization; yet all three processes can occur simultaneously, though their predominance varies in time and space (Lewis and Maund (1976)). The processes of depopulation and suburbanization will now be considered, albeit briefly.

The first half of the nineteenth century was characterized by a relatively slow growth in the rural population of North American and west European countries, from either a net in-migration and natural increase or a natural increase that balanced a net outmigration. With the onset of industrialization from about the 1860s onwards, the drift to the towns began with the consequent loss of population over large parts of the countryside (Lawton, 1968). According to Lord Eversley (1907, p. 280), this rural depopulation was due not only to 'the greater prosperity and general rise of wages in the manufacturing and mining districts', but also to a 'growing disinclination to farm work among labourers in rural districts, to the absence of opportunities in their vocation, and to a desire for the greater independence and freedom of life in towns'. Rural depopulation across much of the countryside continued apace well into the 1920s (Fig. 7.1), when it began to slow down due to a combination of two factors: first, the beginnings of suburbanization by the growth of an 'adventitious' rural population on the fringes of cities; and second, the transformation of certain urban-industrial areas by the economic crises of the period into depressed areas (Zelinsky, 1962; Saville, 1957).

After World War II, this duality in the demographic experiences of the country-side became even more evident. The depopulation of the agricultural and more remote rural regions escalated to such an extent that outmigration was often accompanied by natural decrease (Vince, 1952). Although economic factors continued to be regarded as the cause of this decline, several researchers emphasized the significance of a wider 'socio-economic health' factor within the process (Jones, 1976). What these studies emphasized was a deepening in the decline of the rural infrastructure, resulting in an accelerated downward spiral in local opportunities and facilities (Mitchell, 1950). It is little wonder, therefore, that Hannan (1970, p. 81) concluded the root cause of the young leaving western Ireland was determined by 'beliefs about one's ability to fulfil "economic type" aspirations locally'. Although the majority of these moves were to major cities and metropolitan areas, they often incorporated a distinctive migration system such as described by Hillery and Brown (1965, p. 47) for southern Appalachia; 'its localities did not belong to the same migration system but rather to a collection of 'backyards' connected through migration to non-Appalachian areas, often distant cities'.

From the 1950s onwards, the process of outward movement from cities to the surrounding countryside accelerated within a context of increasing affluence, efficient public transport and a rising rate of private-car ownership. Yet this suburbanization of the countryside was not entirely uniform; for example, Hart and Salisbury (1965) revealed that in the Midwest of the United States there was a strong relationship between distance from a town (up to 25 miles) and village population growth during the 1950s. Although Fuguitt and Field (1972) emphasized the significance of population size and the socio-economic character of villages in the process, essentially these changes allowed an ever growing number of city people to realize their desire to live in the countryside. In the American context, 'the image of green fields, small communities and basic primacy of family relationships may draw people away from the problems of the metropolis to the more manageable world of the fringe' (Pahl, 1965, p. 75). Yet this must not be over-emphasized, since Pahl (1965) and Lewis (1979) have argued that the availability of relatively cheap housing was of considerable significance within the process. Clearly, what has been described here is a process of urban decentralization, effectively transforming more accessible villages into largely middle-class commuter communities; in other words, the beginnings of what has become known as counterurbanization (Ambrose, 1974).

COUNTERURBANIZATION AND THE RURAL TURNAROUND

Definitions

By the 1970s the duality in the demography of the countryside was overtaken by what Berry (1976, p. 24) has proclaimed as 'a turning-point in the American experience. Counterurbanization has replaced urbanization as the dominant force shaping the nation's settlement patterns'. Subsequently, such a process was identified in the majority of other advanced economies and was generally regarded as the principal driving force in the rural population turnaround (Champion, 1989; Cross, 1990; Brown and Wardwell, 1980; Frey, 1987; Bedford, 1983; Joseph *et al.*, 1988).

The precise nature of counterurbanization has been a source of considerable debate; a wide variety of terms such as 'regeneration', 'dispersal', and 'core–periphery migrations' have been used in association with the concept. Perhaps much of the confusion surrounding the study of counterurbanization can be put down to this varied terminology and to imprecise or contradictory conceptual frameworks that accompany it (Zelinsky, 1977; Vartiainen, 1989). According to Berry (1976), counterurbanization essentially involved a change between the extended suburbanization resulting from the expansion of the metropolitan regions that had characterized urban development for three-quarters of a century, and the contemporary phenomenon in which some areas beyond the metropolis exhibited significant population growth.

The same phenomenon was embodied in the notion of a 'clean-break', the term used by Vining and Strauss (1977) to differentiate between the trend of population concentration in urban areas and the trend of population deconcentration into peripheral areas. Where a positive relationship is found between growth and settlement size, Fielding (1982) argued that urbanization is deemed to be the prevailing tendency at work, whereas counterurbanization is identified where the rate of growth becomes greater for places of progressively smaller size. Further more, Robert and Randolph (1983) distinguished between 'decentralization', the growth in the hinterland of an expanding metropolitan region, and 'deconcentration', the movement beyond that region to places further down the urban hierarchy. Yet, Beale (1975) considered counterurbanization simply as a pattern of population change characterized by faster rates of rural and non-metropolitan population growth than those experienced in urban and metropolitan centres, whereas McCarthy and Morrison (1977) identified the trends as nothing more than rural repopulation, namely an upturn in non-metropolitan population change rates and an increase in the non-metropolitan population. According to some authorities, the process by the late 1980s was becoming more an outcome of urban spillover than the reorientation towards country living implicit in the terminology (Maher and Stimson, 1994; Sant and Simons, 1993). However, by the early 1990s it was evident that the geographical nature of counterurbanization involved at least four common factors:

■ Growth was occurring at progressively lower levels of the urban hierarchy.
■ Population increase was spreading through extended suburbanization.
■ Buoyant rates of growth were being recorded outside metropolitan areas, especially in remoter rural areas.
■ Population was shifting from traditional urban industrial areas towards locations more favoured in environmental terms.

Clearly, counterurbanization is a complex and multifaceted process and has to be treated with caution as an agent for understanding the revival in fortunes of the contemporary rural population. So far it has been most widely applied as a description of the redistribution of national population change; on the other hand, its role

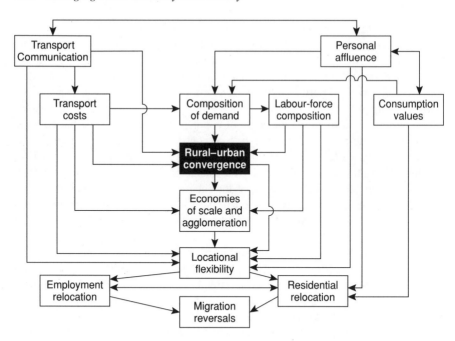

Figure 7.2   A paradigm of the non-metropolitan turnaround (Reprinted from Brown and Wardwell, 1980, by permission of Academic Press Inc., London)

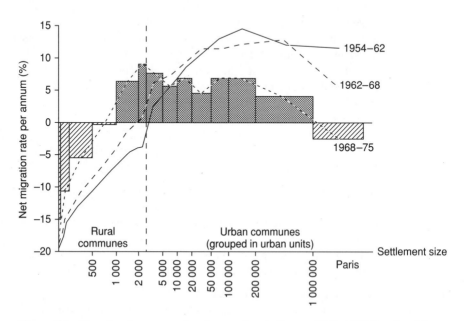

Figure 7.3   Net migration rates and settlement sizes in France, 1954–1975 (Reprinted from Fielding, 1982, by permission of Elsevier Science Ltd.)

in effecting demographic and social change within the countryside is less well understood.

The driving forces

In the ensuing debate about the causes of counterurbanization and the rural turn-around, the principal focus of attention has been upon its hierarchical and regional nature within a context of a reduction in the comparative advantage enjoyed for so long by large cities and the emergence of a 'new' spatial economy (Fielding, 1982). Among the myriad of causal explanations suggested by various authorities, three appear to be of particular relevance (Frey, 1987). The *period explanation* emphas-izes the role of the peculiar economic and demographic circumstances of the 1970s, including the energy crisis and various recessions, and the attenuation of metropolitan growth by the postwar baby boom and a sharp growth in retirees (Kontuly and Bierens, 1989). Clearly, this perspective views the 1970s population redistribution as an aberration and therefore argues that after a short time there will be a return to urban concentration. On the other hand, the *regional restructuring explanation* highlights the significance of new organizations of production and a new spatial division of labour, as well as the shift towards service-based industries, in stimulating a greater spread of activities and population towards smaller places and the rural periphery.

The *deconcentration theory* takes the view that long-standing preferences towards lower density locations are being less constrained by institutional and tech-nological barriers. The changing industrial structure, the rising standards of living, and technical improvements in transport, communication and production are leading to a convergence – across size and place – in the availability of amenities that were previously accessible only in large places (Fig. 7.2). Since the countryside gener-ally possesses a greater range of undeveloped and high amenity opportunities than industrial core regions, this perspective expects an acceleration in the redistribution towards the periphery. From the growing evidence at both regional and local scales, it would appear the deconcentration theory is the most relevant in any attempt to explain the rural turnaround (Champion, 1989).

Emerging trends

The onset of counterurbanization within advanced economies, after its first iden-tification by Beale (1975) in the United States, was regarded by many to be unex-pected and likely to be short-lived (Berry, 1976). Yet this is a rather surprising viewpoint since its antecedents were evident over half a century ago. For example, population growth was taking place on the metropolitan and urban fringe of British cities immediately after World War II, from where it spread into more intermediate locales in the 1960s, then to the rural periphery during the 1970s (Jones *et al.*, 1986). Similarly, Fielding (1982) revealed in France a shift in the net migration rates of settlements of different sizes from the early 1950s to the mid-1970s away from large cities and towards the smaller, rural communities (Fig. 7.3). More specifically, Champion (1987) has argued that counterurbanization had its origins at least in the first half of the 1960s and, contrary to popular belief, the 1970s actually experienced a downward trend in a longer cycle of deconcentration.

Vining and associates (1977, 1978) in a series of influential papers emphasized that counterurbanization was taking place throughout the majority of advanced economies during the 1970s. A clear shift of population over time between core and peripheral regions was occurring in the national space economy; in particular, a reversal in the direction of net population flow in favour of sparsely populated regions or at least a drastic reduction in the level of the net flow. Not surprisingly, this research did not necessarily reveal the existence of a true rural turnaround because the scale of analysis was too broad and there was also a failure to consider the whole of the settlement system.

During the 1970s, numerous studies of counterurbanization in the United States emphasized how it involved a regional shift of population from the traditional areas of growth in the north and east to those of the west and south, as well as significant growth in the non-metropolitan areas, in particular just beyond the metropolitan limits. Consequently, it may be regarded as an extension to the decentralization process (Fuguitt, 1985). Elsewhere, however, the turnaround was considerably more patchy and discontinuous; population growth in Australia during the 1970s was focused primarily outside the commuting hinterland of the capital cities, particularly in the closely settled areas of the growth states of New South Wales, Queensland, and Western Australia (Hugo, 1988). Beyond these areas, growth was more selective and partial in the sense that not all similarly sized places experienced the same trend at the same time (Hugo and Smailes, 1985). Yet there was enough commonality to support the view that deconcentration was prevalent across much of rural Australia (Sant and Simons, 1993; Sorenson and Epps, 1993).

On the other hand, despite the confirmation of a rural revival in most developed countries during the 1970s, the simplistic distinction of metropolitan/non-metropolitan or urban/rural adopted in many studies tended to hide the increasing significance of shifts down the urban hierarchy and across different kinds of settlements. According to Lewis (1992), in England and Wales between 1971 and 1981 there was a distinct difference in the demographic experiences of various types of settlement (Fig. 7.4). Not surprisingly, it was the metropolitan districts which lost population most heavily between 1971 and 1981. In sharp contrast, all the non-metropolitan districts underwent a marked 'turnaround' in their demographic fortunes. Within this diverse group of districts, it was those characterized by medium to small settlements which experienced the greatest amount of population growth by 1981. However, as revealed by Fig. 7.5 (on page 140), much of this non-metropolitan growth was again regionally concentrated; this time in the southern half of England and Wales, particularly East Anglia, the south Midlands, the West Country, and the Welsh Marches.

Unfortunately, in many studies there has been a tendency to overemphasize the widespread significance of the rural turnaround, even in the remotest of rural regions. In England and Wales, several rural districts continued to lose population during the 1970s, and even within districts undergoing sharp increases, some parishes were still experiencing depopulation (Weekley, 1988). Besides that, the

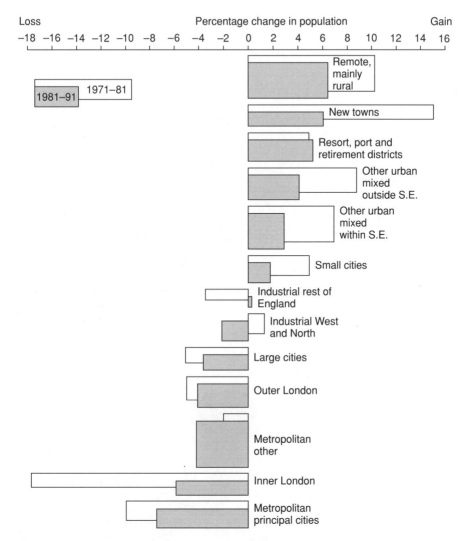

Figure 7.4 Population change in England and Wales by area type: 1971–1981 and 1981–1991

absolute levels in population growth in regions associated with the counterurbanization process were often substantially less than the growth of population numbers in large metropolitan cities. Similarly, at a more local level, as illustrated in the case of Radnorshire in central Wales, the large percentage increases in population between 1971 and 1981 were often only small numerical increases (Fig. 7.1, page 132). All this evidence suggests the rural turnaround was far from uniform, and its numerical significance limited, across the whole of the countryside.

According to White *et al.* (1989, p. 267), 'just as counterurbanization became accepted and population deconcentration trends were be ng extrapolated into the future, the patterns shifted back in the early 1980s, with metropolitan areas once

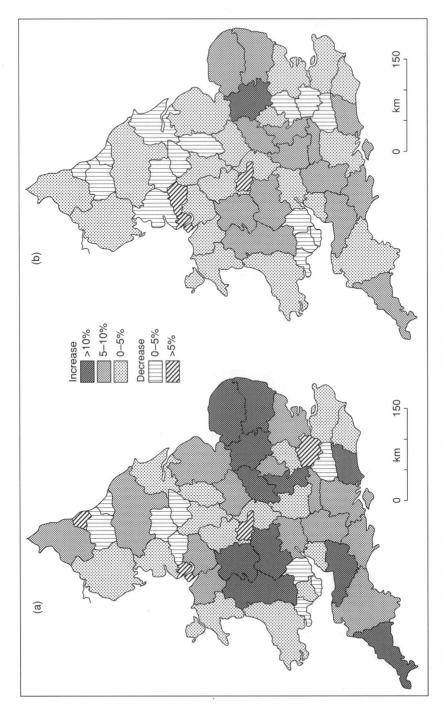

Figure 7.5 Population change in England and Wales by county: (a) 1971–1981 and (b) 1981–1991

again growing faster than non-metropolitan areas'. This 'new' metropolitan revival was evident in the United States, Canada, Ireland and Australia (Frey, 1993; Keddie and Joseph, 1991; Walsh, 1991; Hugo, 1994), whereas in several other countries, counterurbanization continued to be a feature of population redistribution, admittedly at a slower rate (Champion, 1994; Kontuly and Vogelsang, 1988; Serow, 1991). The nature of this slowing down in the turnaround process can be illustrated once again from England and Wales (Fig. 7.4, page 139). Between 1981 and 1991 the metropolitan/non-metropolitan balance had become more diverse at a time when the national population had declined marginally (−0.1%). Among the metropolitan districts, Greater London halved its rate of population loss compared to the 1970s, whereas the rate of loss in the remaining metropolises continued to rise. Once again the non-metropolitan districts continued to increase their populations, albeit at a reduced rate.

Significantly, the 'remoter, mainly rural districts' were the fastest growing type of area after 1981, outpacing all other types in terms of both percentage and absolute size of population increase. Of the 42 districts growing more than 10% in the period 1981–1991, only four were to be found north of the Wash, reaffirming the growth pattern of the 1970s. By the 1980s, the rural growth regions were evolving into a belt extending from East Anglia through Northamptonshire and Oxfordshire to Wiltshire, Dorset and Hampshire (Fig. 7.5). The emergence of this growth belt became more evident by the late 1980s with a shift in growth from the peripheral counties of the South-East to the surrounding counties. Beyond this belt, population growth was more uneven than in the 1970s, being most evident in parts of the South-West, mid-Wales and the Welsh Marches.

Despite the resurgence in metropolitan population growth, the rural turnaround was still clearly evident in the early 1990s; however, it was becoming much more socially and spatially selective, with the location of settlement in its regional context becoming a more significant parameter than trends in the country as a whole (Sherwood, 1986). Nowhere is this more evident than in contemporary Australia (Maher and Stimson, 1994; McKenzie, 1994). By the early 1990s, the metropolitan centres were once again growing faster than the non-metropolitan areas; population growth in the non-metropolitan areas had become even more discontinuous than in the 1970s (Salt, 1992). Non-metropolitan growth was generally focused in the sphere of metropolitan influence or in the rapidly growing coastal strip extending from New South Wales to the Sunshine Coast, north of Brisbane (Sant and Simons, 1993). In contrast, depopulation was becoming more evident in the wheat/sheep belts and the extensive dryland grazing areas (Hugo and Smailes, 1992). Only in the outback did certain regional centres continue to experience a growth in their populations (McKenzie, 1994; Hugo, 1994). Maher and Stimson (1994) likened these regionally biased demographic trends in Australia to those found by Lewis *et al.* (1991) in Britain, where rural population growth was becoming focused primarily within a 60 km semicircular belt around London. In this sense, Maher and Stimson (1994) suggest that counterurbanization is increasingly becoming another type of concentration.

The demographic undulations experienced by developed countries in the 1980s inevitably raise the question of whether they signal the end of the non-metropolitan turnaround (Champion, 1988). If the definition of the rural turnaround is taken to be simply that non-metropolitan growth exceeds metropolitan growth, the early 1980s certainly witnessed the end of the turnaround (Frey, 1988). Such a conclusion, however, needs to be treated with some caution; in particular, the necessity to give greater consideration to the process over a time period greater than just two decades or so. For example, Johnson (1989) has argued that, despite a reversal between 1980 and 1989 in the United States to metropolitan growth rates exceeding those in non-metropolitan areas, when viewed in historical terms, fundamental changes have taken place in the redistribution of its population. Thus non-metropolitan demographic trends for at least 40 years before the 1970s in the United States, as in most developed countries, were those of widespread depopulation involving substantial outmigration, low natural increase, and in some places even natural decrease. Compared with the previous four decades, the 1970s and 1980s in the United States were different in at least three ways. First, non-metropolitan population gains were both much larger and more extensive than in previous decades. Second, both the incidence and magnitude of net immigration to non-metropolitan areas were substantial in the 1970s and also the 1980s, which contrasted with the net outflow of population from non-metropolitan areas in earlier decades. Only those rural districts adjacent to the metropolitan areas gained population during the 1960s. Third, compared with earlier decades, small settlements and the remoter countryside experienced substantial proportional gains in population. Even in counties where, during the 1980s, metropolitan areas were increasing in population at a faster rate than non-metropolitan areas, many of these differences still remained significant. Such a perspective raises doubt over Berry's (1988) claim that excessive attention has been paid to short-term fluctuations at the expense of longer-term trends of 'urbanward migration'. However, before it can be asserted that the rural turnaround is still alive and well, counterurbanization has to be conceived within a much wider perspective than adopted hitherto (Vartiainen, 1989).

THE HOUSEHOLD AND HOUSEHOLD TURNOVER

In the debate over the pervasiveness and long-term future of counterurbanization, it is unfortunate how the majority of studies have overlooked the fact that, as a process, it involves much more than just population growth and net migration gain. If the emphasis is switched from population numbers to changes in household numbers and in the composition of local populations, then a truer picture of the significance of counterurbanization for the economy and society of small settlements becomes more readily apparent.

By conceiving population and migration change in terms of individuals, there has been a failure to realize that a vital component in the process is the household. It is the household that requires housing with all of its ramifications for the local housing market; new household formations have evolved since World War II and their numbers have grown significantly in recent decades. Any analysis of household trends as a component in the counterurbanization process needs to be conceived within a context of a national decrease in the average size of household during the past three decades. More specifically, the mean household size in England and Wales has fallen from 3.26 in 1961 to 2.46 in 1990. In 1961, one in seven households comprised just one person; by 1990 it was one in four (Lewis, 1992). By 2010 it is predicted that England as a whole will have an extra 2.8 million households – an increase of 15% – and much of this growth will be concentrated in the southern half of the country. Such trends will have immense consequences for housing and rural life there.

The significance of the decline in household size in determining population trends in rural settlements can be illustrated as follows. If the population of a village remained the same (say 1000) between 1961 and 1990, but the average size of its households followed the national trend, there would be an additional 101 households by 1990 (407 compared with 306 in 1961) and, therefore, a need for 55 more dwellings to house them. Alternatively, the population of this village would have to decrease from 1000 in 1961 to 752 before it could experience any numerical loss of households or housing units. This relationship between population change and household change has implications for the supply of housing and is fundamental to a proper understanding of the processes of demographic change. It suggests that, in situations of overall decline, the rate of population decrease will be greater than the rate of household decrease; whereas in situations of overall growth, the rate of household increase will be greater than the rate of population increase. Furthermore, it is quite possible for a settlement to experience population losses at the same time as increases in the number of households and housing stock; the reverse situation is less likely since this will imply an increase in household size.

Figure 7.6 (overleaf) reveals some of these issues with reference to Breckland, a rural district in eastern England. Between 1961 and 1971, Breckland was still experiencing net losses both of population and households, though the figure was less than 1% for household losses. Even so, 28% of the parishes gained in population and as many as 43% gained in households, particularly those with larger settlements. What is clearly evident from Fig. 7.6 is a pattern of diminishing household losses and increasing gains as one moves up the settlement hierarchy, even though these gains were not as great as in the previous decade. These changes could be viewed as the beginnings of the 'turnaround'.

Like other rural areas, Breckland experienced a sharp renaissance in its demographic fortunes during the 1970s, although its population increase (8.3%) must not be overemphasized since a substantial proportion of its parishes (45%) continued to lose population, particularly the smaller ones. Between 1971 and 1981 there was also an increase of 14% in the district's households and only about a third of its parishes experienced losses. Although these losses tended to be in small parishes, note that, even within the smallest settlement category, more parishes were gaining households than losing them. In other words, on the basis of household changes, the counterurbanization of the 1970s was much deeper and more pervasive than is revealed by population change alone. These trends were reinforced during the 1980s. This example illustrates the necessity of examining demographic change by household as well as by population in a study of the rural turnaround.

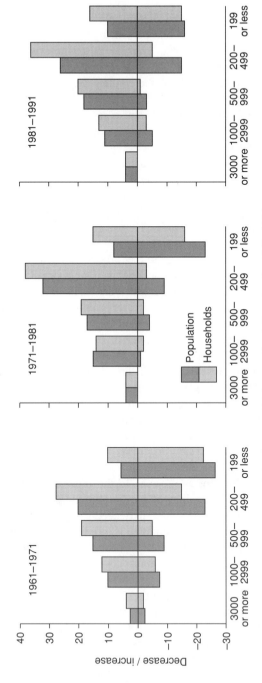

Figure 7.6 Population and household change by parish in Breckland, East Anglia, 1961–1991

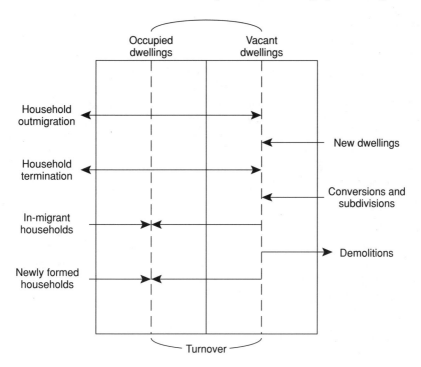

Figure 7.7 The process of household turnover

In effecting a recomposition of rural communities, the underlying mechanism involved is again not necessarily population growth and net migration but rather the character of the flows into and out of the rural settlements (Lewis *et al.*, 1991). At a regional level, it is now realized that net migration provides only a partial picture of the process of population redistribution since net migratory gains or losses reveal only surface ripples of powerful cross-currents. So Rogers (1990) has gone as far as to claim that interregional migration and its effect on demographic change can only really be understood if the migrant inflows and outflows are considered separately. In other words, 'there are no net migrants; instead there are people who are arriving at places or leaving them. Why they are doing so is central to understanding the dynamics of growth and decline' (Morrison, 1977, p. 61).

The significance of differential population flows into and out of the countryside as a determinant of population recomposition is even more evident at the local scale (Lewis *et al.*, 1991). The ebb and flow of individuals and households into and out of communities, often called *turnover*, is the basic mechanism involved in the process. However, regardless of the characteristics of the in-migrant to a community, the move can only take place by the provision of opportunities in the housing stock. As Fig. 7.7 shows, housing vacancies can occur in two major ways: through additional dwellings and through the creation of vacancies in the existing dwelling stock by household migration or household termination. The migrants can therefore increase their relative strength over the resident population within a community not only by the occupation of new housing but also by their replacement within the existing housing stock. Where migrants differ in their socio-economic characteristics from those whom they replace, what may be termed 'replacement selectivity', then a recomposition of the population will take place. Regardless of general tendencies to growth or decline in a community's household population, it is the characteristics of incomer households in comparison with the resident households which form the basis of social change. But turnover can be a means for both changing and maintaining the population characteristics of a community. Thus high turnover need not necessarily produce social change if the replacement selectivity is low; but if turnover and replacement selectivity are both high as well as population growth, there is little doubt that social change will be widespread within the countryside.

This suggests that, in the debate over the future of the counterurbanization process and its impact upon rural society, it is the household and the changing composition of the population that are more meaningful than population growth alone, and their combination is usually a more effective guide. From this perspective there is little doubt that counterurbanization is still prevalent in the contemporary countryside, so the next section will investigate its underlying mechanisms.

A BEHAVIOURAL PERSPECTIVE

In order to gain a better understanding of why and how people move into and out of the countryside, several researchers of the counterurbanization process have advocated the need to initiate more microstudies of the process since there has recently been an overemphasis upon both the spatial analytic approach (e.g. Boyle, 1995) and the more critical perspective (e.g. Marsden *et al.*, 1993; Murdoch and Marsden, 1994). These perspectives provide little insight into the behaviour of individual migrants, except so far as such behaviour can be deterministically predicted by reference to changes in capitalist production or such variables as distance and town size.

As long ago as 1970, Hagerstrand (1970, p. 8) made the telling point that 'nothing truly general can be said about aggregate regulants until it has been made clear how far they remain invariant with structural differences at the micro level'. Thus the context of this micro approach emphasizes the fact that 'people at different places and at different positions in the social structure have different degrees of knowledge about, different perceptions of, and are able to benefit to different degrees from, opportunities at places other than those in which they currently reside' (Walmsley and Lewis, 1993, p. 169). However, this does not imply that migrants are sovereign decision makers who simply maximize some subjective notion of desirability. Nothing could be further from the truth, because all potential migrants are constrained to some extent by a wide range of variables that include economic power, social and emotional ties, information availability, job opportunities and dwelling requirements. Thus, microscale behavioural studies of migration need to take account of both choice in decision making (the demand side of migration opportunities) and constraints and restrictions, especially within the labour and housing markets.

Despite these pleas from researchers for more microstudies of contemporary rural migrations, there are still relatively few. Part of the reason for this dearth of studies lies in the perennial problem involved in 'asking questions' – whether highly structured or qualitative – and the inherent difficulty of linking macro and micro analyses (Cadwallader, 1989). Despite these difficulties, a behavioural perspective does provide a framework for raising some significant questions:

- How do people decide to move?
- Why do migrants choose to move?
- What constraints affected the migration decision?
- Who are the migrants?
- Where do people migrate to?
- In what ways do past migration experiences influence contemporary decisions to migrate?

Decision to move and spatial search

If a microscale behavioural approach is adopted in the study of migration, the focus of attention falls on the decision-making processes of the actors involved. This in turn is often differentiated into three separate activities:

- The decision to leave the present residence and move to a new one.
- The search for a suitable location.
- The selection of a new residence from among those that are suitable.

Residence in this sense incorporates both the dwelling itself and the location of that dwelling relative to features of the environment (e.g. work, school, shops) that are important to the decision maker.

The majority of the residential decision-making models (e.g. Brown and Moore, 1970; Brown, 1983) were conceptualized in terms of a stress-satisfaction formulation derived from Wolpert's (1965) concept of place-utility. Briefly, this viewpoint argues that the basis of any decision to migrate is the belief that the level of satisfaction obtained elsewhere is greater than the present level of satisfaction. The difference between the two levels can therefore be regarded as a measure of stress (Clark and Cadwallader, 1973). The major emphases within these models are the nature of motivation, the preferences underlying a decision to move, and examination of the search and evaluation process.

Not all decisions are free and a wide variety of movement constraints operate within all societies; consequently, several researchers find these models relatively unrealistic (Marsden *et al.*, 1993). In response to such criticism, De Jong and Fawcett (1979) hypothesized that migration is the result of (1) the strength of the value expectancy derived from the intentions to move, (2) the direct influences of background individual and aggregate factors, and (3) the potential modifying effects of often unanticipated constraints and facilitations which may intervene between intentions and actual behaviour. To appreciate the outcome of these processes and the nature of the constraints, it is necessary to understand the reasons behind migration.

Both economic and non-economic reasons have been advanced as prime motives generating contemporary migrations in the countryside, though their relative significance varies with the geographical scale of the study and the nature of the localities involved (Williams, 1981). According to Fuguitt (1985, p. 269), for example, 'both employment growth in non-metropolitan America as well as preferences for smaller places and attractive environmental settings underlie the turnaround'. However, Williams and McMillan (1980) argued this was a too simplistic and restrictive interpretation of the decision of households relocating to non-metropolitan America because it is vital to distinguish between those households that simultaneously chose a destination, given a certain reason for leaving (simultaneous decisions), from those employing different criteria for each stage of the process. It was evident that environmental factors such as landscape, climate and community feeling were particularly significant among all households in determining the reason for leaving a place of origin. The principal criterion in choosing a new destination was often a household's ties with a place, often known as location-specific capital, and involving former residence on the part of the migrants, previous holiday experiences, and the location of friends and relatives. But remember that over 25% of the migrants left their previous place of residence for employment reasons, and over two-thirds of that 25% stated how the availability of a job had influenced their choice of destination.

Several local studies of rural migration decision making in contrasting environments have revealed similarities and differences with rural America (Harper, 1991; Halfacree, 1994). For example, the reasons given for movements to amenity areas, such as the coastal region of northern New South Wales, were much more specifically related to the nature of the local environment (Walmsley *et al.*, 1995). By far the most significant factors involved in attracting migrants to the area were climate, lifestyle and environment, although improved job and dwelling opportunities were also mentioned. On the other hand, among the long-distance migrants to rural districts in lowland England, employment and retirement factors were the prime determinants for leaving their former residence (Lewis and Sherwood, 1994). In the selection of new destinations, these migrants indicated not only the significance of locational ties but also employment opportunities and a series of environmental factors. Moves within the district, however, were driven by personal, housing and environmental factors, whereas not surprisingly, the choice of individual village or small town was determined by housing factors (Fig. 7.8).

This evidence indicates how the rural turnaround does not reflect entirely a desire to replace an urban lifestyle with a rural one. In parts of the countryside, whether an amenity area or an urban commuting zone, there is difficulty in ascertaining whether population shifts are truly counterurban instead of merely a residential trade-off between house prices, lifestyle and accessibility to work. Nevertheless, the existence of employment-related factors in all localities indicates the continuing significance of economic restructuring in the turnaround process. Above all, any

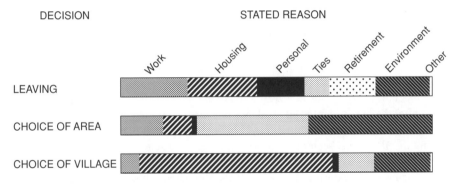

Figure 7.8 Motives for migration: a summary compiled from four districts in lowland England, 1993

understanding of the motives underlying contemporary migrations must distinguish not only between in- and outmigrants but also between local and regional moves.

The rural turnaround is a good illustration of a decision-making situation where there are many different channels of information, and where a good deal of effort is put into the acquisition of information, much of it spatial in nature. Thus the residential choice process can be conceived as operating within the context of the employment and housing market, with migratory moves being the result of the interaction between the migrant and the markets. This interaction involves search (Preston, 1987). The concept of *search* itself encompasses several interrelated characteristics: it is a goal-directed activity; it involves a complex process of information gathering; it takes place in a context of uncertainty; a point is reached where search ends and a choice is made; and it takes place within a set of constraints (Gale *et al.*, 1990). The process of searching may therefore be characterized by its duration, the number and type of information services used, the number of places searched, the number of houses examined, and the radius of the area searched (Golledge *et al.*, 1995).

Yet, despite the significance of search in the migration process, there have been surprisingly few attempts to elucidate its significance within the rural environment (Huff, 1986). From the limited evidence, however, it appears the search procedure adopted by rural migrants is both similar to and different from the procedure used by urban migrants. According to Walmsley *et al.* (1995), the number of locations considered by migrants to the coastal region of northern New South Wales was as few as three regions; a further six or so localities were searched within the chosen region. Lewis and Sherwood (1994) in their study in lowland England suggested that a distinction has to be made in the search process between those households with locational ties to a district and those without. Households with locational ties had usually made their choice of district well before the decision to migrate was activated. Not surprisingly, those households without a connection to their eventual chosen district tended to consider some eight or nine possibilities, using a variety of information sources, though the final choice was very much based on personal experience.

For both groups, however, the choice of village or small town involved a tendency to focus their search on part of a district; and the availability of a particular type of house was the most significant criterion. However, access to schools and an urban centre was also of importance for those households with young children. Significantly, it was also revealed that the search procedures adopted by potential migrants during the 1980s involved a relatively short period of search, particularly short after the area of preferred residence had been identified; this often led to many secondary moves after the initial move. In contrast, search in the rural environment during the 1990s has been much more drawn out and selective, with little evidence of secondary moves. No doubt this reflects the depressed state of the economy and the housing market during the early 1990s compared with earlier periods.

Underlying most of these findings is the suggestion that there is more to residential satisfaction than simply employment and housing requirements. In other words, individuals identify with locations, including their residential location, and endow them with social meaning. For example, Feldman (1990) applied the concept of self-identity to explain the cognitive and behavioural disposition towards the city, the suburb and the village, and suggested that it influenced future residential intentions. For many researchers, moves to the countryside during recent decades reflect a latent disposition towards rural living, a kind of 'rural idyll' so beloved by the media. However, several studies have emphasized that underlying many of the 'reasons' given for migrating to the countryside was a deep-seated desire for a more *rural* residential environment (Halfacree, 1994; Lewis *et al.*, 1991). However, from the available evidence this must not be overemphasized, since identity in some social settings might have effects on behaviour opposite to its effects in other settings.

Migration selectivity and social change

It has been well established that some persons migrate, often frequently, to new places of residence whereas others remain behind. Rather than the entire population, migration tends to characterize a limited segment of a population who make frequent and repeated moves (Lind, 1969). Essentially, this mover–stayer dichotomy suggests that the inducements to migrate do not exert their force equally, hence the tendency for some individuals to be more prone to migrate than others (Lewis, 1982).

Within the counterurbanization literature, much attention has been given to the socio-economic characteristics of the migrants involved. Apart from the initial overemphasis on the so-called 'urban dropout', there is much evidence to support the view that the process is dominated by the middle classes, whether they be retirees or long-distance commuters. However, as counter-urbanization has spread and deepened within the countryside, it has become more evident that the middle classes themselves contain significantly different fractions of households (Murdoch, 1995). For example, Cloke and Thrift (1987) and Phillips (1993) have emphasized the growing significance of the service classes in effecting intraclass differences in rural Britain, whereas in rural New Zealand the so-called rural lifestylers and retired farmer household were pre-eminent in several village populations (Joseph and Chalmers, 1992). Yet numerous studies across the rural United States and rural Australia indicate the socio-economic background of the migrants was much more diverse than suggested in rural Britain and elsewhere, thus emphasizing a greater range of employment and environmental opportunities (Brown and Wardwell, 1980; Hugo and Smailes, 1985). This is most evident in the coastal region of New South Wales, where recent migrants included not only professional households but also tradespeople and unskilled workers (Walmsley *et al.*, 1995).

Unfortunately, within the counterurbanization literature there has been an over-emphasis on the characteristics of the long-distance migrant, thus overlooking the fact that contemporary demographic and social changes within the countryside have to be conceived within the context of the whole migration process. For example, among the villages and small towns of lowland England, Lewis and Sherwood (1994) were able to identify several different migrant characteristics according to the distances moved. On a regional scale, three distinct types of migrant households were revealed: (1) late-age professional families without children; (2) early-age service families with children; and (3) retired households. Yet, among these households, the counterurban migrant should not be overemphasized, since among the first two groups were a number of households that had moved from other rural districts.

Among the early-age service families there was a tendency to seek employment in a town and to choose to live in its rural hinterland. This group had particularly high turnover rates, since their migration was related to their career, either at the bidding of their employer or while changing jobs. Significantly, for many of these 'spiralists', residence in the countryside was part and parcel of their residential decision making. Locally there was also considerable movement, a kind of sifting and sorting of households, which was creating a considerable degree of differentiation among the villages and small towns (Murdoch *et al.*, 1994). Generally, during the 1980s and early 1990s there was a tendency for the retired and the late-age professional households to move towards the smaller and more picturesque villages, whereas the young service families and the lower income households were more likely to be found in the larger and more accessible villages and small towns. Both regional migrations and local movements are clearly effecting considerable demographic and social recomposition within rural society.

Life cycle and household formation

A major factor in determining mobility is the changing threshold of residential dissatisfaction that accompanies changes in the family life cycle. From formation to dissolution of the cycle, critical events can be identified which increase or decrease the propensity to migrate (e.g. marriage, birth of children, last child leaving home and retirement). Despite several early researchers confirming the significance of the life cycle in the migration process, later writers have revealed that life changes are never perfectly correlated with mobility levels; in particular, they fail to involve changing career patterns, income and social status (Sandefur and Scott, 1980; Lewis,

1982). But probably the most significant reason to revise the family life cycle has been the diversification of family and non-family structures and behaviour over the past few decades (Grundy, 1985). These new household formations, such as the growth in single-parent households, young and old single-adult households, and two earners in a dual-headed household, have led Stapledon (1980) to suggest an expanded life-cycle model, emphasizing that a variety of households now exist which tend to fluctuate in time and space. Essentially three questions arise from these changes:

- How do new household formations influence the residential decision-making process?
- How do the transitions from one household formation to another initiate migration?
- Do different household formations locate themselves in different localities?

Not surprisingly, these questions have stimulated considerable interest, particularly the changing role of women in the migration process (Boyle and Halfacree, 1995; Halfacree, 1995). Within a rural context, Lewis and Sherwood (1994) found that, in 'shire' England, divorce, widowhood and the onset of adulthood were a signficant stimulant to migration, especially moves to the cities, or at least to the local small town or large village. Not unexpectedly, it was also found that the mobility rates of single-headed family households were much higher than those of dual-headed family households. Of the single-headed family households, 90% were female with children, largely having been created by marriage dissolution, and often the resultant fall in income and entry into the labour market compelled them to seek cheaper accommodation in locations that provided better access to jobs and childcare facilities. Inevitably, destinations of female-headed households were away from villages and remoter settlements towards more accessible locations and urban areas.

As a result of greater female participation in the labour market, it has also been suggested that the balance of power within the residential decision-making process has begun to shift away from the employed male (Bonney and Love, 1991). In a study of upper middle-class families in the Ile de France, Fagnani (1993, p. 176) suggested how the 'presence of two earners within a dual-headed household potentially both increases and narrows residential location options, as well as increasing the influence of the female in the residential decision'. Significantly, the more equal the spouses' incomes, the more equal their respective influences in the process. However, as a result of cultural dictates linked to the image of being a 'good mother', of scheduling which does not recognize two-earner couples, and time–space constraints involved in rural living, women face contradictory realities and must cope with conflicting obligations.

In the Ile de France (Fagnani, 1993) and lowland England (Lewis and Sherwood, 1994) it was found that the most salient factors influencing the residential choice of dual-headed households were proximity to the workplace, access to public transport, shopping, schools and daycare centres for those children below three years of age. Considerable negotiations were involved in this process and compromise solutions were sought. Generally, it was identified that women played a major role in the decision to reside in the countryside, or to choose another location in the countryside, because of womens' family commitments to the home. Location of the home represents a more important aspect of a woman's life than of her partner's. It was most likely, therefore, that a couple's residential choice was determined by the location of the wife's workplace (Madden, 1981). The emergence of these new household formations and their different residential requirements is now playing a significant role in determining residential mobility in the countryside.

Life courses and residence history analysis

In order to gain a deeper understanding of the migration process in the countryside, recent studies have begun to adopt a longitudinal approach – residence history analysis within a life-course context (Fielding, 1989; Warnes, 1992). From a migration viewpoint this approach achieves three things:

- Identification of the circumstances of the household at the time of the migration.
- Placing of the particular migration within the life and migration history of the household.
- Identification of the possible influence of previous migratory experiences upon the present, and likely future, moves.

Essentially, residence history analysis involves the recording of the previous place of residence of migrants over their life history within the context of the events or transitions that occur as people age (Pryor, 1979; Warnes, 1986; Bailey, 1989). These transitions can be interpreted within the context of changing historical conditions. The focus is upon the biological, socio-demographic, psychological and socially constructed basis of change through the life course and the pathways through the various structures in the major role domain of life (Rossi, 1985). The nature of a life-course perspective has been most succinctly summarized by Hareven and Adams (1982, p. 86) as 'the interrelationships between individual and collective family behaviour as they constantly change over people's lives and in the context of historical conditions. The life-course approach is concerned with the movement of individuals over their own lives and through historical time, with the relationships of family members to each other as they travel through personal and historical time'.

Following the pioneering work of Price (1963) on the residential history of immigrants in Australia, and Hagerstrand's (1975) study of lifetime moves in rural communities in Sweden, contemporary work has begun to focus on issues related to the counterurbanization process. For example, Mooney (1991), using a biographical method, compiled a record of a number of individuals' changing housing circumstances in order to identify those factors influencing changes in housing needs and the problems encountered when attempting to change a housing situation. On the other hand, Lewis and Sherwood (1994) employed a residential history analysis to trace the place experience of migrants over their lives. What they concluded was that the majority of the in-migrants, including those from metropolitan areas, had considerable experience of rural living before their last move. Even among the long-distance migrants, over 70% had experienced rural living sometime during their lives, whereas nearly 60% of all migrants had resided at some time in at least two different rural areas. Within this rural experience, it was evident that return migration, particularly among the retired, was of increasing significance; on the other hand, it questions the validity of the picture often created by the media of the counterurban townie with little or no rural experience and detached from the realities of rural living. From these two casestudies, it is evident that a life-course approach to residential history shifts the emphasis of the analysis from the migration experience of place to the place of experience of migrants, as well as providing real-life experiences of people as they negotiate their way through the housing and settlement system (Forrest and Murie, 1987).

CONCLUSION

This chapter has sought to examine the process of migration and socio-demographic change in the countryside with a particular focus on the past 30 years or so. From the wide-ranging evidence presented here, three issues are worthy of some further consideration: the role of a behavioural approach in the analysis of the rural turn-around, the impact of migration on local communities, and the significance of the definition adopted for counterurbanization and the rural turnaround in any assessment of likely future trends.

In 1980 Clark and Moore (1980, p. 18) argued that 'if faith is placed in analytical studies of the individual without regard to societal relations which define the context within which relocation adjustments are made, one is liable to design programmes and policies whose outcomes fail to live up to expectations'. During the past decade, therefore, behavioural research has emphasized the point that migration takes place within the context not only of imperfect knowledge but also of considerable constraint. These constraints reflect the structured nature of society and manifest themselves locally in terms of differential access to employment and housing. However, despite considerable research on the nature and operation of the institutions involved in the provision of employment and housing, little attention has been given to how they directly influence the decisions made by individual households as to when and where to move. The increased adoption of a biographical approach and the conception of local migrations within regulation theory provide a means for potentially overcoming this impasse (Halfacree and Boyle, 1993; Goodwin et al., 1995).

Although in this chapter migration has been conceived as a response to social, political, economic and cultural changes in society, it is also quite clear that migration can also be viewed as an independent variable since it can initiate change itself (Lewis, 1982). Apart from the demographic and compositional changes it has engendered among rural communities, migration has also effected changes in their economies, housing market, service provision and local power relations. In other words, there is much evidence to suggest these migrations have at times accentuated the underlying problems of rural living for a number of rural dwellers (ACORA, 1990; Cloke, 1995). For example, the young still leave the countryside in search of employment, whereas newcomers have accentuated the difficulties of the low income households in gaining access to adequate housing. But note that even the migrants themselves can experience difficulties in their new localities, such as the elderly's need for an effective and accessible medical provision (Joseph and Chalmers, 1992). The continuing talk of a rural renaissance by academics and the sanitization of the countryside by the media do nothing more than cloud the still widespread and persistent deprivation in the countryside (Chapter 11).

Finally, it has been emphasized that the unidirectional view adopted in many studies of the rural turnaround gives only a partial picture of the recomposition of rural communities. In other words, in order to unravel the complexities of the process, it is necessary to look beyond numerical increase and net migration gains and to focus upon the selectivity of migrants within the context of household turnover. From such a viewpoint, it would appear that in several countries counterurbanization is deepening despite some attenuation in rural population growth during the late 1980s. By widening the conceptualization of the process, a different view on the turnaround–reversal debate may be forthcoming. In order to interpret this broader conceptualization of the counterurbanization process, a more specific time–space perspective is required. Champion (1988) has gone some way towards achieving this by suggesting the need to consider the interplay of three sets of factors at a particular point in time: *deconcentration*, including better transport and communication, the dispersal of educational and related services, and the growth of residential and recreational opportunities; *concentration*, including the advantages for those economic activities requiring substantial access to other national and international activities best located in large cities; and *demographic factors*, including age- or cohort-specific preferences for different residential settings. Depending upon the form of these preferences, their impact would be to accentuate the overall pattern towards either concentration or deconcentration. In many advanced economies, deconcentration remained predominant into the early 1990s and there is growing evidence to suggest it will continue into the next century.

REFERENCES

ACORA (1990) *Faith in the countryside*, Report of the Archbishops' Commission on Rural Areas. Churchman Publishing, Worthing.

Ambrose, P. (1974) *The quiet revolution*. Chatto and Windus, London.

Bailey, A.J. (1989) Getting on your bike: what difference does a migration history make? *Tijdschrift voor Economische en Sociale Geografie*, **80**, 312–17.

Beale, C.L. (1975) *The revival of population growth in non-metropolitan America*. Economic Research Service, US Department of Agriculture, ERS-605.

Bedford, R.D. (1983) Repopulation of the countryside. In Bedford, R.D. and Sturman, A.P. (eds). *Canterbury at the crossroads: issues for the eighties*, Special Publication 8. New Zealand Geographical Society, Christchurch, pp. 277–306.

Berry, B.J.L. (ed) (1976) *Urbanisation and counterurbanisation*. Sage, Beverly Hills CA.

Berry, B.J.L. (1988) Migration reversals in perspective: the long-wave evidence. *International Regional Science Review*, **11**, 245–60.

Bonney, N. and Love, J. (1991) Gender and migration: geographical mobility and the wife's sacrifice. *Sociological Review*, **39**, 335–48.

Boyle, P.J. (1995) Rural in-migration in England and Wales, 1980–81. *Journal of Rural Studies*, **11**, 65–78.

Boyle, P.J. and Halfacree, K.H. (1995) Service class migration in England and Wales, 1980–81: Identifying gender-specific mobility patterns. *Regional Studies*, **29**, 43–58.

Brown, D.L. and Wardwell, J.W. (eds) (1980) *New directions in urban–rural migration: the population turnaround in rural America.* Academic Press, New York.

Brown, J.M. (1983) The structure of motives for moving: a multidimensional model of residential mobility. *Environment and Planning A*, **15**, 1531–44.

Brown, L.A. and Moore, E.G. (1970) The intra-urban migration process: a perspective. *Geografiska Annaler B*, **52**, 1–13.

Cadwallader, M. (1989) A conceptual framework for analysing migration behaviour in the developed world. *Progress in Human Geography*, **13**, 494–511.

Champion, A.G. (1987) Recent changes in the pace of population deconcentration in Britain. *Geoforum*, **18**, 379–407.

Champion, A.G. (1988) The reversal of migration turnaround: resumption of traditional trends? *International Regional Science Review*, **11**, 253–60.

Champion, A.G. (ed) (1989) *Counterurbanization: the changing place of population deconcentration.* Edward Arnold, London.

Champion, A.G. (1994) Population change and migration in Britain since 1981: evidence for continuing deconcentration. *Environment and Planning A*, **7**, 1501–20.

Clark, W.A.V. and Cadwallader, M. (1973) Locational stress and residential mobility. *Environment and Behaviour*, **5**, 29–41.

Clark, W.A.V. and Moore, E.G. (eds) (1980) *Residential mobility and public policy.* Sage, New York.

Cloke, P. (1995) Rural poverty and the welfare state: discursive transformation in Britain and the USA. *Environment and Planning A*, **27**, 1001–16.

Cloke, P.J. and Thrift, N. (1987) Intra-class conflict in rural areas. *Journal of Rural Studies*, **3**, 321–33.

Cross, D. (1990) *Counterurbanization in England and Wales.* Avebury, Aldershot.

De Jong, G.F. and Fawcett, J.T. (1979) Motivations and migration: an assessment and value expectancy research model. In *Workshop in Microlevel Approaches to Migration Decisions*, East–West Centre, Honolulu, pp. 48–64.

Eversley, Lord (1907) The decline in the number of agricultural labourers in Great Britain. *Journal of the Royal Statistical Society*, **70**, 267–319.

Fagnani, J. (1993) Life course and space: dual careers and residential mobility among upper-middle class families in the Ile-de-France region. In Katz, C. and Monk, J. (eds) *Full circles geography of women over the life course.* Routledge, London, pp. 171–87.

Feldman, R.M. (1990) Settlement identity: psychological bond with home plans in a mobile society. *Environment and Behaviour*, **22**, 183–229.

Fielding, A.J. (1982) Counterurbanization in western Europe. *Progress in Planning*, **17**, 1–52.

Fielding, A.J. (1989) Inter-regional migration and social change: a study of south-east England based upon data from the Longitudinal Study. *Transactions of the Institute of British Geographers*, **14**, 24–36.

Forrest, R. and Murie, A. (1987) The affluent home owner: labour market position and the shaping of housing histories. *Sociological Review*, **35**, 370–403.

Frey, W.H. (1987) Migration and depopulation of the metropolis: regional restructuring or rural renaissance? *American Sociological Review*, **52**, 240–57.

Frey, W.H. (1988) The re-emergence of core regions growth: a return to the metropolis? *International Regional Science Review*, **11**, 261–67.

Frey, W.H. (1993) The new urban revival in the United States. *Urban Studies*, **30**, 741–74.

Fuguitt, G.V. (1985) The non-metropolitan turnaround. *Annual Review of Sociology*, **11**, 259–80.

Fuguitt, G.V. and Field, D.R. (1972) Some population characteristics of villages differentiated by size, location and growth. *Demography*, **9**, 295–308.

Gale, N., Golledge, R.G., Halperin, W.C. and Couclelis, H. (1990) Exploring spatial familiarity. *Professional Geographer*, **42**, 299–313.

Goldstein, S. (1976) Facets of redistribution: research challenges and opportunities. *Demography*, **13**, 34–43.

Golledge, R.G., Dougherty, V. and Bell, S. (1995) Acquiring spatial knowledge: survey versus route-based knowledge in unfamiliar environments. *Annals of the Association of American Geographers*, **85**, 134–58.

Goodwin, M., Cloke, P. and Milbourne, P. (1995) Regulation theory and rural research: theorising contemporary change. *Environment and Planning A*, **27**, 1245–60.

Grundy, E. (1985) Divorce, widow-hood, remarriage and geographic mobility among women. *Journal of Biosocial Science*, **17**, 415–35.

Hagerstrand, T. (1970) What about people in regional science? *Papers and Proceedings of the Regional Science Association*, **24**, 7–21.

Hagerstrand, T. (1975) On the definition of migration. In Jones, E. (ed) *Readings in social geography*. Oxford University Press, Oxford, pp. 200–209.

Halfacree, K.H. (1994) The importance of the rural in the constitution of counterurbanization: evidence from England in the 1980s. *Sociologia Ruralis*, **34**, 164–89.

Halfacree, K.H. (1995) Household migration and the structuralisation of patriarchy: evidence from the USA. *Progress in Human Geography*, **19**, 159–82.

Halfacree, K. and Boyle, P. (1993) The challenge facing migration research: the case for a biographical approach. *Progress in Human Geography*, **17**, 333–48.

Hannan, D.F. (1970) *Rural exodus*. Chapman, London.

Hareven, T. and Adams, K.J. (eds) (1982) *Ageing and life course transitions: an interdisciplinary perspective*. Guildford Press, New York.

Harper, S. (1991) People moving to the countryside: case-studies of decision making. In Champion, A.G. and Watkins, C. (eds) *People in the countryside*. Paul Chapman, London.

Hart, J.F. and Salisbury, N.E. (1965) Population change in middle western villages: a statistical approach. *Annals of the Association of American Geographers*, **55**, 140–60.

Hillery, G.A. and Brown, J.S. (1965) Migrational systems of the southern Appalachians: some demographic observations. *Rural Sociology*, **30**, 47.

Huff, J.O. (1986) Geographic regularities in residential search behaviour. *Annals of the Association of American Geographers*, **76**, 208–27.

Hugo, G.J. (1988) *Australia's changing population*. Oxford University Press, Melbourne.

Hugo, G.J. (1994) The turnaround in Australia: some first observations from the 1991 census. *Australian Geographer*, **25**, 1–17.

Hugo, G.J. and Smailes, P.J. (1985) Urban–rural migration in Australia: a process view of the turnaround. *Journal of Rural Studies*, **1**, 11–30.

Hugo, G.J. and Smailes, P.J. (1992) Population dynamics in rural Australia, *Journal of Rural Studies*, **8**, 29–52.

Johnson, K.M. (1989) Recent population redistribution trends in non-metropolitan America. *Rural Sociology*, **54**, 301–26.

Jones, H., Caird, J., Berry, W. and Dewhurst, J. (1986) Peripheral counterurbanization: findings from an integration of census and survey data in northern Scotland. *Regional Studies*, **20**, 15–26.

Jones, H.R. (1976) The structure of the migration process: findings from a growth point in mid-Wales. *Transactions of the Institute of British Geographers*, **NS2**, 421–32.

Joseph, A.E. and Chalmers, L. (1992) Servicing seniors in rural areas: an issue for New Zealand. *Proceedings of the New Zealand Geographical Society and Institute of Australian Geographers Inaugural Joint Conference*, Auckland, pp. 393–402.

Joseph, A.E., Keddie, P.D. and Smit, B. (1988) Unravelling the population turnaround in rural Canada. *Canadian Geographer*, **32**, 17–30.

Keddie, P.D. and Joseph, A.E. (1991) The turnaround of the turnaround? Rural population change in Canada 1976–86. *Canadian Geographer*, **35**, 367–79.

Kontuly, T. and Bierens, H.J. (1990) Testing the recession theory as an explanation for the migration turnaround. *Environment and Planning A*, **22**, 253–70.

Kontuly, T. and Vogelsang, R. (1988) Explanations for the intensification of counter-urbanization in the Federal Republic of Germany. *Professional Geographer*, **40**, 42–54.

Lawton, R. (1968) Population changes in England and Wales in the later nineteenth century: an analysis of trends by registration districts. *Transactions of the Institute of British Geographers*, **44**, 55–74.

Lewis, G.J. (1979) *Rural communities: a social geography*. David and Charles, Newton Abbot.

Lewis, G.J. (1982) *Human migration: a geographical perspective*. Croom Helm, London.

Lewis, G.J. (1992) Counterurbanization and social change in rural Britain: some recent trends. *Netherlands Geographical Studies*, **153**, 61–73.

Lewis, G.J. and Maund, D.J. (1976) The urbanisation of the countryside: a framework for analysis. *Geografiska Annaler B*, **58**, 17–27.

Lewis, G.J. and Sherwood, K.B. (1994) *Rural mobility and housing*. Working Papers 7–10, Department of Geography, University of Leicester.

Lewis, G.J., McDermott, P. and Sherwood, K.B. (1991) The counterurbanization process: demographic restructuring and policy response in rural England. *Sociologia Ruralis*, **31**, 309–20.

Lind, H. (1969) Internal migration in Britain. In Jackson, J.A. (ed) *Migration*. Cambridge University Press, Cambridge.

McCarthy, K.F. and Morrison, P.A. (1977) The changing demographic and economic structure of non-metropolitan areas in the United States. *International Regional Science Review*, **2**, 123–42.

McKenzie, F. (1994) *Regional population decline in Australia*. Australian Government Publishing Service, Canberra.

Madden, J. (1981) Why women work closer to home. *Urban Studies*, **18**, 181–94.

Maher, C.A. and Stimson, R.J. (1994) *Regional population growth in Australia*. Australian Government Publishing Service, Canberra.

Marsden, M.T., Murdoch, J.H., Love, P., Munton, R. and Flynn, A. (1993) *Constructing the countryside*. UCL Press, London.

Mitchell, G.D. (1950) Depopulation and rural social structure. *Sociological Review*, **42**, 11–24.

Mooney, E. (1991) Access to rural housing in Scotland: some preliminary observations from Argyll and Bute. In Lewis, G.J. and Sherwood, K.B. (eds) *Rural mobility and housing*, Proceedings of the Institute of British Geographers Rural Study Group Annual Conference, Leicester, pp. 57–77.

Morrison, P. (1977) The functions and dynamics of the migration process. In Brown, A.A. and E. Neuberger (eds) *Internal migration: a comparative perspective*. Academic Press, New York, pp. 61–72.

Murdoch, J. (1995) Middle-class territory? Some remarks on the use of class analysis in rural studies. *Environment and Planning A*, **27**, 1213–30.

Murdoch, J. and Marsden, T. (1994) *Reconstituting rurality: the changing countryside in an urban context*. UCL Press, London.

Murdoch, S.H., Hwang, S.-S. and Hoque, H.N. (1994) Non-metropolitan residential mobility revisited. *Rural Sociology*, **59**, 236–54.

Pahl, R.E. (1965) *Urbs in rure: the metropolitan fringe in Hertfordshire*. Geographical Paper 2, London School of Economics.

Pahl, R.E. (1966) The rural/urban continuum. *Sociologia Ruralis*, **6**, 299–329.

Parker, R. (1970) *The common stream*. Penguin, Harmondsworth.

Phillips, M. (1993) Rural gentrification and the processes of class colonisation. *Journal of Rural Studies*, **9**, 123–40.

Preston, V. (1987) Spatial choice behaviour in urban environments. *Urban Geography*, **8**, 374–79.

Price, C. (1963) *Southern Europeans in Australia*. Oxford University Press, Melbourne.

Pryor, R.J. (ed) (1979) *Residence history analysis*. Studies in Migration and Urbanization 3, Department of Demography, Australian National University, Canberra.

Robert, S. and Randolph, W. (1983) Beyond decentralization: the evolution of population distribution in England and Wales, 1961–81. *Geoforum*, **14**, 75–102.

Rogers, A. (1990) Requiem for the net migrant. *Geographical Analysis*, **22**, 283–300.

Rossi, A. (ed) (1985) *Gender and the life course*. Aldine, New York.

Rowland, D.T. (1979) *Internal migration in Australia*. Australian Bureau of Statistics, Canberra.

Salt, B. (1992) *Population movements in non-metropolitan Australia*. Australian Government Publishing Service, Canberra.

Sandefur, G.D. and Scott, W.J. (1981) A dynamic analysis of migration: an assessment of the effects of age, family and career variables. *Demography*, **18**, 355–68.

Sant, M. and Simons, P. (1993) The conceptual basis of counterurbanization: critique and development. *Australian Geographical Studies*, **31**, 113–26.

Saville, J. (1957) *Rural depopulation in England and Wales 1851–1951*. Routledge & Kegan Paul, London.

Serow, W.J. (1991) Recent trends and future prospects for urban–rural migration in Europe. *Sociologia Ruralis*, **31**, 269–80.

Sherwood, K.B. (1986) Social settlement structures in rural Northamptonshire. *Cambria*, **13**, 41–61.

Sorenson, T. and Epps, R. (eds) (1993) *Prospects and policies for rural Australia*. Longman Cheshire, Melbourne.

Stapledon, C.M. (1980) Reformulation of the family life-cycle concept: complications for residential mobility. *Environment and Planning A*, **12**, 1103–18.

Vartiainen, P. (1989) Counterurbanization: a challenge for socio-theoretical geography. *Journal of Rural Studies*, **5**, 217–25.

Vince, S.W.E. (1952) Reflections on the structure and distribution of rural population in England and Wales, 1921–31. *Transactions of the Institute of British Geographers*, **48**, 173–87.

Vining, D.R. and Kontuly, T. (1978) Population dispersal from major metropolitan regions. *International Regional Science Review*, **3**, 49–73.

Vining, D.R. and Strauss, A. (1977) A demonstration that the current deconcentration of population in the United States is a clean break with the past. *Environment and Planning A*, **9**, 751–58.

Walmsley, D.J. and Lewis, G.J. (1993) *People and environment: behavioural approaches in human geography*. Longman, London.

Walmsley, D.J., Epps, W.R. and Duncan, C.J. (1995) *The New South Wales North Coast 1986–1991: who moved where, why and with what effect?* Australian Government Publishing Service, Canberra.

Walsh, J.A. (1991) The turnaround of the turnaround in the population of the Republic of Ireland. *Irish Geography*, **24**, 117–25.

Wardwell, J.M. (1977) Equilibrium and change in non-metropolitan growth. *Rural Sociology*, **42**, 156–79.

Warnes, A.M. (1986) The residential mobility histories of parents and children, and relationships to present proximity and social integration. *Environment and Planning A*, **18**, 1581–94.

Warnes, A.M. (1992) Migration and the life course. In Champion, A.G. and Fielding, A.J. (eds) *Migration processes and patterns*, vol I, *Research progress and prospects*. Belhaven, London, pp. 175–82.

Weekley, I.G. (1988) Rural depopulation and counterurbanization: a paradox. *Area*, **20**, 127–34.

White, S.B., Osterman, J.D. and Binkley, L.S. (1989) *The sources of suburban job growth*. Urban Research Centre, University of Wisconsin, Milwaukee WI.

Williams, S. (1981) The nonchanging determinants of non-metropolitan migration. *Rural Sociology*, **46**, 183–202.

Williams, J.D. and McMillan, D.B. (1980) Migration decision-making among non-metropolitan bound migrants. In Brown, D.L. and Wardwell, J.M. (eds) *New directions in urban–rural migration*. Academic Press, New York, 189–211.

Wolpert, J. (1965) Behavioural aspects of the decision to migrate. *Papers and Proceedings of the Regional Science Association*, **15**, 159–72.

Zelinsky, W. (1962) Changes in the geographic patterns of rural population in the United States 1790–1960. *Geographical Review*, **52**, 492–524.

Zelinsky, W. (1971) The hypothesis of the mobility transition. *Geographical Review*, **61**, 219–49.

Zelinsky, W. (1977). Coping with the migration turnaround: the theoretical challenge. *International Regional Science Review*, **2**, 125–27.

CHAPTER 8

RURAL INDUSTRIALIZATION
David North

THE CONTEXT OF GLOBAL RESTRUCTURING

The past quarter-century has seen a dramatic change in the location of manufacturing industries at the global scale which has taken the form of a shift of
capacity and employment away from the older industrial economies of western
Europe and the United States towards Japan and some of the developing market
economies, particularly those of the Pacific Rim. For example, the United States'
share of world manufacturing production decreased from 40% in 1963 to 24%
in 1987, whereas for Japan it increased from 5% to 19% over the same period
(Dicken, 1992, p. 23). Moreover, countries such as South Korea and Taiwan
have been experiencing rapid manufacturing growth since the 1960s (with annual
growth rates in excess of 10% during the 1970s and 1980s), whereas countries
like France and the United Kingdom have struggled to achieve any growth in
manufacturing output.

For the United Kingdom, the decline in the competitiveness of its manufacturing
base and the associated change in the locus of production globally have become
manifest in a relentless tide of factory closures and job losses, which have affected
the main industrial cities and regions of the country over the past two decades.
From its peak of 8.5 million workers in 1966, manufacturing in the United Kingdom has now shrunk to around 4.3 million. Another indication of the deindustrialization of the United Kingdom is the mounting trade deficit in manufactured
goods from 1982–83 onwards, reaching £15 billion by the end of the 1980s. Thus
the United Kingdom, along with other developed economies, has become more and
more dependent upon service sectors rather than manufacturing; manufacturing
now accounts for only one-fifth of total employment and just over one-fifth of the
domestic product. In analysing the phenomenon of rural industrialization within
Europe and the United States, therefore, it is important to bear in mind that the
overall context is one of deindustrialization at the level of national economies.

As well as this global shift in the location of manufacturing, there has been a
second major change which has been affecting the character and organization of
manufacturing industry and in turn its locational requirements. This is the trend

161

away from the mass production of standardized products towards more flexible production systems, often known as the shift from Fordism to neo-Fordism (or post-Fordism) (Allen and Massey, 1988). Advances in production technology (such as computer-aided design and manufacture) and information technology have greatly assisted this trend by making it possible for firms to produce in small batches without any significant increase in operating costs. Associated with these techno-logical changes have been changes in consumer tastes and expectations in favour of more differentiated products, leading to more customized patterns of consump-tion and the proliferation of niche markets. It is argued (e.g. Piore and Sabel, 1984; Scott and Storper, 1988) that these developments are resulting in new flexible production systems which are characterized by greater vertical disintegration in the ways in which firms are organized, smaller production units, and increased subcon-tracting of activities to other firms, thereby creating new market opportunities for smaller businesses (Shutt and Whittington, 1987).

Increased flexibility in the use of labour, both in terms of numerical flexib-ility (such as through the increasing use of part-time workers and outworkers) and functional flexibility (such as through the reduction of skill demarcations), is another characteristic of the shift towards neo-Fordist production methods. Given the radical nature of these changes in the organization of production, it is hardly surprising they also have profound implications for the location of manufacturing activities. Consequently, the areas associated with the mass-production industries of the Fordist era are losing out to the 'new industrial spaces' which are emerging under neo-Fordism. Any interpretation of rural industrialization should be set within the context of these broader-level changes.

This chapter aims to examine the evidence for the urban–rural shift which has occurred in the location of industry within developed economies over the past three decades and to consider the various explanations proposed by different writers. The focus will be primarily upon manufacturing industry, although it is recognized that the distinction between manufacturing and services is becoming increasingly anachronistic in the context of the information-based economy of the twenty-first century. After considering the evidence for rural industrialization in western Europe and the United States, the chapter will examine the evidence relat-ing to the United Kingdom in some detail before summarizing five alternative explanations for the urban–rural shift which reflect different theoretical traditions within economic geography. The rest of the chapter focuses upon the contribution of two rather different types of enterprise to rural industrialization. Firstly, the important contribution of small businesses to rural industrialization will be ana-lysed, looking both at urban and rural contrasts in new business formation and in the growth of established firms. Secondly, the role played by large national and multinational corporations in the form of branch plant investment in rural regions will be discussed. A key aspect will be the extent to which rural industrialization in the form of the growth of small businesses on the one hand, and branch plant investment on the other hand, can be interpreted satisfactorily using the theories outlined earlier in the chapter.

Table 8.1 The urban–rural shift in the European Community

Type of region	Change in percent share of GDP 1970–1977	Change in percent share of industrial output 1970–1977	Change in manufacturing employment 1973–1979
Highly urbanized regions (21)	−1.7	−2.9	−6.0
Urbanized regions (23)	−0.3	+0.2	−6.9
Less urbanized regions (32)	+0.9	+1.1	−1.6
Rural regions (30)	+1.1	+1.6	−2.9

Source: Keeble *et al.* (1983), based on Eurostat data.

THE EVIDENCE FOR RURAL INDUSTRIALIZATION

The urban–rural shift in western Europe and the United States

One of the most significant trends in the location of manufacturing industry within the developed economies of western Europe and the United States since the 1960s has been its growth in rural areas relative to urban areas, a phenomenon which has become known as the urban–rural shift (Fothergill and Gudgin, 1982). This spatial shift has therefore been superimposed upon the underlying deindustrialization of these national economies. However, the shift is not unique to these more advanced economies as it also appears in some recently industrializing countries such as Greece, Spain, Portugal and the Irish Republic.

An analysis of the regional performance of manufacturing during the 1970s within the nine member countries of the European Community at that time leads Keeble and his colleagues to the conclusion that 'the urban–rural shift is remarkable for its scale, its consistency with regard to settlement size, and its equal incidence under different national economic conditions' (Keeble *et al.*, 1983). When regions were classified according to the degree of urbanization (based on population density and the proportion of the population living in urban agglomerations of 100 000 inhabitants or more), the rural regions were shown to have performed favourably in terms of all three measures used, i.e. gross domestic product, industrial output, and manufacturing employment (Table 8.1). Over the relatively short period 1970–1977, the rural regions together with the less urbanized regions increased their share of the European Community's economic activity at the expense of the urbanized and highly urbanized regions by no less than two percentage points.

Although all types of region experienced a decline in manufacturing employment over the period, the rate of decline in rural regions was less than half that experienced by the urbanized regions. More detailed comparisons of individual countries showed that the shift has been equally true for more recently industrializing economies, such as Italy, as for older industrial economies, such as West

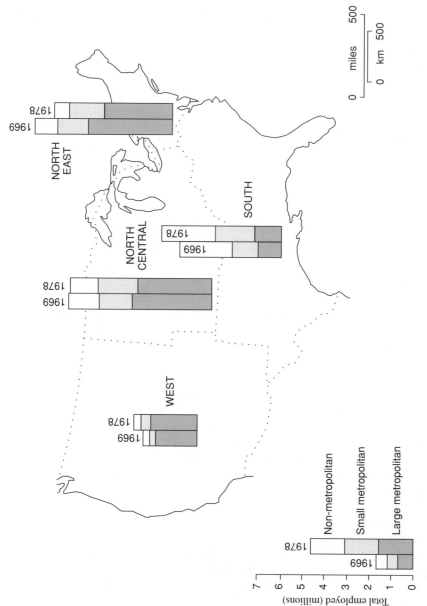

Figure 8.1 United States metropolitan and non-metropolitan maufacturing employment by region, 1969 and 1978

Germany and France. The urban–rural differences are especially remarkable, given that the urban regions have relatively favourable structures compared with the bias towards declining sectors in the rural regions. As Keeble and his colleagues note, sectoral trends alone would suggest a rural–urban shift rather than the urban–rural shift which has occurred.

In the United States, the first signs of a reversal in the centralization of manufacturing employment in the major cities were identified in the 1950s, but the decentralization process gathered pace during the 1960s and 1970s (Lonsdale, 1985). According to Lonsdale, there was a 2.3 million increase in non-metropolitan manufacturing employment between 1954 and 1978, the national share in these areas increasing from 22% to 29%. During the 1960s, when US manufacturing employment as a whole increased by 20%, non-metropolitan areas achieved a 31% increase compared to a 15% increase in metropolitan areas; and during the 1970s, when the trend had reversed in metropolitan areas with manufacturing employment falling by 3.5%, non-metropolitan areas still achieved a 12% growth (Lonsdale and Seyler, 1979). There has, however, been some questioning of these results on methodological grounds.

Estall (1983) has shown how a rather different perspective on the urban–rural shift of manufacturing in the United States arises when changes in the boundaries of the standard metropolitan statistical areas (SMSAs) are taken into account. He shows that the larger part of the non-metropolitan growth of manufacturing during the 1970s occurred in those areas which were adjacent to established urban areas, not in more truly rural areas, representing a continuation of existing trends towards decentralization and suburbanization rather than a sharp break with the traditional locational pattern of industrial investment. Moreover, the urban–rural shift has been greatly influenced by what has happened in the southern states, since it is here that manufacturing employment has been growing and where a substantial proportion has traditionally been in non-metropolitan counties (Fig. 8.1). In comparison, there is little evidence of an urban–rural shift in the other three main regions of the United States. There are therefore some important reasons for questioning whether the urban–rural shift of manufacturing in the United States has been as significant as some writers have suggested.

Some further evidence in support of Estall's view is provided in a study of the decentralization of high technology industry in the United States (Barkley, 1988). Over the period 1975–1982 non-metropolitan employment in high technology industries increased (by 15.5%) but not as rapidly as for metropolitan areas (28.2%). However, the performance of the non-metropolitan counties relative to their metropolitan counterparts did improve when account was taken of their sectoral composition and how different sectors were performing at the national level (using shift share analysis, see below). It was then found there had been a small shift of high technology employment away from the largest metropolitan areas; counties adjacent to the metropolitan areas and those further away both benefited. As with US manufacturing employment as a whole, it was the counties (both metropolitan and non-metropolitan) in the South and West regions which were the main beneficiaries from the trend, and the urban centres in the North which were the main losers.

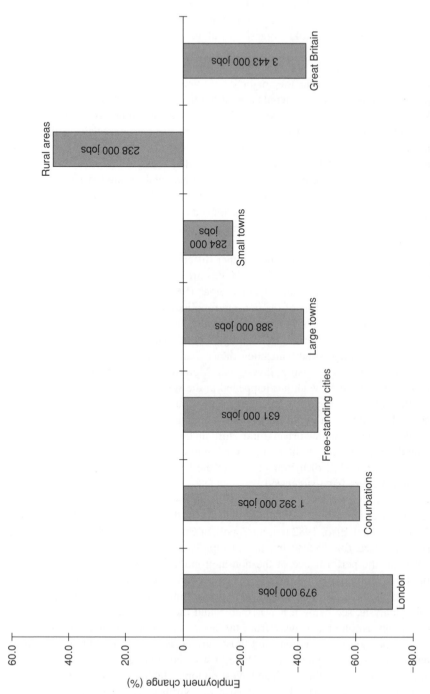

Figure 8.2 Employment change in manufacturing 1960–1991 by urban–rural category (Compiled from Department of Employment statistics)

The study also found that the urban–rural shift tended to be strongest in the case of the chemicals and non-electrical machinery industries, and weakest for electronics and instruments production. Significantly, Barkley also produces evidence to indicate it has been the more mature and slower-growing high technology industries which have shifted to the remoter rural counties, whereas the younger, more innovative sectors have shifted to those rural areas adjacent to metropolitan areas. On this evidence, the urban–rural shift of manufacturing in the United States would appear to have been sectorally differentiated between types of rural area and to have had fewer benefits for the long-term economic development of the remoter rural areas.

The urban–rural shift in the United Kingdom

In order to examine more closely the degree and nature of this urban–rural shift in manufacturing employment, the particularly well-documented case of the United Kingdom will be examined where the evidence spans a period of some 30 years. Figure 8.2 shows the magnitude of the urban–rural shift in Britain over the period 1960–1991. At a time when Britain as whole lost nearly 43% of its manufacturing employment and the major conurbations lost in excess of 60% of theirs, rural areas experienced a 45% increase in manufacturing jobs. By 1991 the rural areas of Britain had almost 250 000 more manufacturing jobs than they had three decades earlier (Gudgin, 1995).

The evidence for an urban–rural shift first surfaced in the work of Fothergill and Gudgin (1982) and Keeble (1980), based on the analysis of manufacturing employment trends during the 1960s and 1970s. Working on data at the local authority district scale, these authors showed a clear inverse relationship between changes in manufacturing employment and settlement size over this period, with the conurbations (and worst of all London) experiencing the sharpest decline in manufacturing employment at one extreme, and rural areas (defined as districts with settlements of less than 35 000 people) an absolute increase in manufacturing employment at the other. The scale of the shift is well illustrated by the fact that, whereas rural areas only had two manufacturing jobs for every five in London in 1960, by 1981 the total manufacturing employment of rural areas was slightly greater than that of London (Fothergill *et al.*, 1985b).

Given the heterogeneity that exists between rural areas, it is necessary to ask whether certain types of rural area have benefited more from the growth of manufacturing employment than others. For the period 1960–1981, Hodge and Monk (1987) have shown that rural areas in southern England (and particularly East Anglia) experienced a more substantial increase than rural areas in northern England, a 32% increase compared with 10%. They have also shown that, at the intraregional scale, it has been rural areas furthest from conurbations which have experienced the greatest increase in manufacturing employment; thus, rural areas more than 80 kilometres from the nearest conurbation experienced a 38% increase compared with a 4% increase for those areas which were less than 40 kilometres away. However, an attempt to correlate both absolute and percentage change in manufacturing employment over the period 1974–1981 with an index of rurality (based on eight indicators of population and occcupational structure and 'peripherality' in 1981) did not yield any significant results, leading the authors to conclude there is a complex relationship between rurality and manufacturing employment, and it cannot easily be reduced to a few generalizations.

Although use of the term *shift* might suggest the movement of firms from urban to rural areas, it is important to realize that the relocation of firms has not been the main component of change in the United Kingdom. Relocation has played its part, especially the movement of firms from London to other parts of the South-East and to East Anglia, but over the United Kindom as a whole the main component has been the differential growth (i.e. the *in situ* expansion and contraction) of existing firms in urban and rural areas (Fothergill *et al.*, 1985a).

The intensity of the shift depends upon the level of new investment and resulting employment trends within manufacturing as a whole. It might be expected, therefore, that the shift would be much weaker or even non-existent during the 1980s, a period when the process of deindustrialization accelerated, with manufacturing investment in real terms below that of the previous decade and manufacturing employment declining by 15% over the period 1981–1989. It is particularly significant that, despite these adverse conditions, the urban–rural shift in manufacturing employment continued during the 1980s, although admittedly not at the same pace as before. Townsend (1993) has shown that manufacturing employment in remoter rural areas continued to increase by 4.6% over the period 1981–1989 compared to a decline in all other types of location, the worst being inner London (–36.2%) and other principal cities (–31.6%).

In fact, the relative shift of manufacturing employment towards rural areas was even greater when allowance was made for the industrial composition of the different types of location using a shift-share analysis, an accepted technique for disaggregating employment change for an area into its national (i.e. what would have happened if manufacturing employment had changed at the national rate), structural (i.e. the expected change given each area's sectoral mix), and differential components (i.e. a residual element which is interpreted as reflecting local advantages and disadvantages) (Table 8.2). In southern Britain, remoter areas were shown to have gained nearly 40 000 more manufacturing jobs over the period 1981–1989 than would have been expected on the basis of national trends and the industrial mix of these areas; whereas at the opposite extreme, inner London lost 80 000 and outer London 90 000 more jobs than expected. Similarly, remoter rural areas in northern Britain gained 33 000 more manufacturing jobs whereas principal cities lost 79 000 more than expected. This more recent evidence not only demonstrates that the urban–rural shift in manufacturing employment has continued during a period which has seen a marked decline of manufacturing employment nationally, but also that the shift has been independent of industrial structure.

Although there has tended to be an urban–rural shift in the employment of most manufacturing industries, it has been particularly evident in the case of high technology industries (defined as synthetic materials, pharmaceuticals, electronic consumer and capital goods, scientific and medical instruments, and telecommunications).

Table 8.2 The urban–rural shift in manufacturing employment 1981–1989 based on shift-share analysis

Type of area	Total change 1981–1989		Differential shift 1981–1989	
	Thousands	Percentage	Thousands	Percentage
South				
Inner London	−102.2	−36.2	−79.7	−28.2
Outer London	−138.0	−34.4	−90.9	−22.7
Non-metropolitan cities	−92.4	−22.2	−16.0	−3.8
Urban and mixed				
urban rural	−43.0	−6.2	+25.2	+3.6
Remoter, mainly rural	+9.8	+3.6	+39.7	+14.5
Total	−433.5	−15.5	−93.0	−3.3
North				
Principal cities	−191.4	−31.6	−78.6	−13.0
Non-metropolitan cities	−42.1	−14.4	+2.5	+0.8
Urban and mixed				
urban rural	−12.3	−5.9	+14.1	+6.8
Remoter, mainly rural	+9.9	+6.2	+33.2	+20.8
Total	−477.7	−14.7	+93.0	+2.9

Source: Townsend (1993) based on Census of Employment Data (NOMIS).

Table 8.3 The urban–rural shift of hi-tech industry in the United Kingdom in the 1980s

County type	Employment in full-time equivalents			
	1981 (000s)	1989 (000s)	Change (000s)	1981–1989 (%)
Greater London	218.5	177.6	−40.9	−18.7
Conurbations (8)	475.3	399.1	−76.2	−16.0
More urbanized counties (14)	319.7	308.9	−10.8	−3.4
Less urbanized counties (21)	375.5	401.8	+26.3	+7.0
Rural counties (20)	91.6	102.7	+11.0	+12.0

Source: Keeble (1992), based on unpublished census of employment statistics from NOMIS.

Thus, over the period 1981–1989, Keeble has shown a 12% increase in employment in high technology industries in rural counties, compared with a 16% decline in the conurbations; the growth is most noticeable in East Anglia, the South-West, central and northern Wales; the decline is greatest in London (Table 8.3) (Keeble, 1992). However, there are two important caveats which emphasize the need to place the urban–rural shift of high technology industry in perspective. The first is that, even at the end of the decade, there were almost three times as many jobs in high technology industries in the rest of the South-East (i.e. the South-East minus

London) than there were in any other region. Thus, although the most rapid growth of high technology employment has been in rural areas, the bulk of employment has remained in more urbanized locations. And second, even within these rural regions, the growth of high technology manufacturing has tended to be concentrated in a particular area; a study of new firms in East Anglia found that three-quarters of the high technology firms were located within 48 kilometres of Cambridge (Gould and Keeble, 1984). This evidence from the United Kingdom is consistent with evidence from the United States cited earlier in the chapter; it shows that the growth of high technology industries has tended to be concentrated in rural areas adjacent to metropolitan areas.

EXPLANATIONS FOR THE URBAN–RURAL SHIFT

Several plausible explanations have been proposed to explain the urban–rural shift in manufacturing employment. Different researchers have their own favourite explanation, which then becomes a kind of leitmotif in all their subsequent writings. To a large extent, these explanations reflect different theoretical traditions within economic geography, i.e. the neoclassical, behavioural and structural analytical frameworks. Rather than some explanations being right and others wrong, it might be better to think of them as indicating the number and complexity of the processes involved, some being more influential at certain times and in certain situations than others. The following is a brief summary of five principal explanations.

The constrained location hypothesis   This is the explanation favoured by Fothergill and Gudgin, who were among the first to identify the urban–rural shift (Fothergill and Gudgin, 1982). Their focus is on the industrial space needs of firms and the abilities of both urban and rural locations to meet them. Two interrelated tendencies were considered important. First there are increases in labour productivity as a result of capital intensification leading to a reduction in employment densities (i.e. the number of workers per given area of factory floorspace). These increases have been affecting firms in both urban and rural areas, but urban areas have tended to find themselves unable to offset them by adding to their stock of industrial floorspace; this constitutes the second tendency. Fothergill and Gudgin have argued that urban areas have found it increasingly difficult to provide both the quantity and the quality of space (e.g. single-storey factories with plenty of space for loading/unloading and parking) required by expanding firms, whereas small towns and rural areas have been able to provide new spacious units as well as the land needed to extend existing units. Thus, reduced employment densities combined with physical constraints have led to a decline of manufacturing employment in urban areas, whereas the availability of land for industrial expansion in rural areas has more than compensated for the fall in employment densities.

The production cost hypothesis   Other writers, notably Tyler *et al.* (1988), have analysed the urban–rural shift in terms of spatial variations in production costs, especially wage costs and land/property costs. Using the kind of logic associated with neoclassical location theory, it is argued that decision takers will pursue optimization strategies which aim to locate their industrial investment in those areas where costs can be minimized and profits maximized. In comparison with urban areas, where high density office development has forced up land rents, many rural areas are seen as low cost locations and therefore attractive to new, space-demanding manufacturing investment. Most of the empirical evidence used by Tyler *et al.* relates to distances from conurbations rather than comparing urban and rural areas as such, but it does demonstrate a clear downward cost gradient with increasing distance from the centre of London for most industries. For example, it was found that gross profits were boosted by 20% if a location was 160 kilometres to the west of London, and by 30% over 320 kilometres away in the South-West region. A similar, although shallower, cost gradient was found in the case of several (but not all) northern conurbations. The main cost differences related to the salary costs of higher-level staff, followed by differences in property rents and rates.

The filter down hypothesis   A somewhat more holistic view of the role played by spatial differences in production costs is provided by those who interpret the decentralization of industrial activity within the context of product life cycles. Much of this work relates to the United States (e.g. Erickson, 1976; Markusen, 1985). This theory argues that, in the early and innovative stages of the product life cycle, urban locations will be favoured because they are capable of providing the skilled labour force, scientific and engineering know-how, and business support services which are needed. However, as other firms enter the industry and competition intensifies, pressure on profits will increase and methods of achieving cost reduction will assume greater importance. Moreover, production becomes more routinized and firms become less dependent upon concentrations of skilled labour. It is during this mature phase that firms will decentralize their more routine activities to rural areas in an effort to reduce production costs, leading to a spatial separation of production activities from administrative and research and development functions.

The capital restructuring hypothesis   There is some overlap between the filter down theory and the capital restructuring theory based on a Marxist or structuralist perspective. Writers such as Massey (1984) have interpreted the urban–rural shift within the context of the major restructuring of production since the 1960s and the attempts by capital (especially in the form of large firms) to restore the balance of class forces in its favour. The driving force is seen as the process of capital accumulation which exploits existing geographical differences arising from previous phases of accumulation, producing new areas of industrial production as well as leading to the demise of previous centres of production. This approach emphasizes the importance of the labour factor in the location of production and the spatial differentiation of labour; to quote Storper and Walker (1984, pp. 3–4), 'as capital develops its capability of locating more freely with respect to most commodity sources and markets, it can afford to be more attuned to labour force differences'.

It is argued, therefore, that the restructuring of production, based on advances in technology and production methods, has loosened the ties which firms had to particular concentrations of skilled labour in the cities and conurbations, and it has opened up new possibilities in areas which did not have a tradition of manufacturing. Smaller towns and rural areas have been favoured, not only because of their lower wage costs, but also because of their lower levels of both unionization and worker militancy. Moreover, some larger firms have been able to gain a virtual monopoly position over the labour force in these rural labour markets. Thus the growth of manu-facturing industry in rural areas has been interpreted as the exploitation by firms (particularly larger firms) of the captive workforce (particularly female labour) to be found in areas whose economy traditionally revolved around agriculture.

Advocates of the capital restructuring interpretation of the urban–rural shift tend to argue that rural areas are playing a particular role within the changing spatial division of labour. In particular, it is argued that it tends to be the more routine, less technically advanced, assembly type functions requiring largely semiskilled workers which are drawn to rural and small town locations rather than those functions which require highly technical and skilled workers (e.g. research and development) or head office, decision-taking functions. In other words, it is claimed that it tends to be the lower levels of the technical division-of-labour hierarchy which have been drawn towards rural locations.

The residential preference hypothesis   This explanation focuses more on entre-preneurial behaviour and the role of qualitative influences on the business location decision. This proves to be a consistent thread in much of Keeble's work and as such has tended to become associated principally with the location of new and small businesses (e.g. Keeble, 1993). It is argued that the founders of businesses are drawn to rural areas by what they perceive as the superior quality of life (including better housing and recreational opportunities) compared to urban areas. In a study of new firm formation in East Anglia, Gould and Keeble (1984) have argued that a high proportion of the entrepreneurs were primarily influenced by environmental considerations in their decision to set up in a rural county, some of them migrating with this intention from urban areas such as London.

In fact, several researchers have seen a clear interrelationship between social changes taking place in rural areas associated with an in-migration of population on the one hand, mainly attracted by quality-of-life considerations, and economic development associated with new business formation on the other hand (e.g. West-head and Moyes, 1992). Thrift (1987, p. 79) has explained the urban–rural shift in terms of 'the predilection of the service class (comprising professionals and man-agers) for the rural idyll' and 'a cascading movement of the service class out of the conurbations and the large cities to the smaller towns and rural areas on their fringes'. It is the service class from which a large proportion of entrepreneurs is likely to come, particularly those with the management experience and educational background needed to create successful businesses. Thrift therefore sees the select-ive nature of the social migration process as being crucial to understanding the urban–rural shift of industry.

Summary   These are the principal explanations that have been put forward for the urban–rural shift. They are not the only ones; other factors may be recognized in different contexts, including aspects of government policy. In the United States, differential tax rates are considered to have encouraged the industrialization of rural areas, as have aggressive recruiting campaigns by both state and local governments (Lonsdale, 1985). In his analysis of the urban–rural shift in the location of manufacturing industry in the United States, Lonsdale identifies as many as nine factors, including labour cost differences; the 'right-to-work' laws in Southern and Plains states, which make it difficult for workers to organize; major improvements in highway and trucking facilities, reducing the needs for industrial agglomeration; and the deteriorating business and social environment found within the major cities (Lonsdale, 1985). Rather than coming down in favour of any single explanation, it is arguably more helpful to adopt an eclectic approach, recognizing that the appropriateness of different explanations varies according to the industrial sector and type of firm being studied. Thus the kinds of processes affecting the development of small businesses in rural areas are likely to be rather different from those behind the decisions of multinational corporations to establish branch plants in rural areas. These are the two different forms of rural industrialization which form the focus of attention for the rest of this chapter.

THE GROWTH OF SMALL BUSINESSES IN RURAL AREAS

A discussion of the contribution that new and small businesses have been making to the economic development of rural areas is justified given 'the overwhelming importance of locally-founded small and medium sized firms, not large externally-controlled firms, in the rural industrialisation process in the UK' (Keeble and Tyler, 1995, p. 977). The encouragement of new firm formation and the growth of existing small businesses have also been given a high priority in rural economic development, as illustrated in the United Kingdom by the Rural Development Commission's Strategy for the 1990s (Rural Development Commission, 1993; North and Smallbone, 1993).

Before highlighting some of the main findings of recent research on rural small and medium-sized enterprises (SMEs), it is important to have some understanding of the changing role that small businesses are playing within national economies, particularly SMEs engaged in manufacturing. As might be expected, the importance of small manufacturing businesses does vary considerably between different Western economies. Rather surprisingly, small manufacturing businesses provide a much higher proportion of manufacturing jobs in the European Union than in the United States; firms employing fewer than 100 workers accounted for 42% of total manufacturing employment in the 12 EU countries taken together in 1988, compared with 20% in the United States (Storey, 1994, p. 25). There are major differences within the EU, with the relative importance of small manufacturing firms tending to be inversely related to the level of economic development; thus small businesses are much more important in countries like Greece and Portugal than in more developed economies like Germany and the United Kingdom. The distribution of manufacturing employment according to enterprise size in the United Kingdom is in fact closer to that of the United States than to the EU average, with 24% of manufacturing employment in firms employing less than 100 workers.

Where the United Kingdom stands out compared with other Western economies is in terms of the resurgence of SMEs that has occurred from the late 1960s onwards. With the possible exceptions of Italy and France, other countries have not shown a clear trend for the importance of small businesses to increase. There is general agreement that for the United Kingdom the overall significance of small businesses, measured in terms of the number of firms, increased during the 1980s; for example, VAT business registration statistics show a 33% increase in the number of registered businesses between 1979 and 1990 (Employment Department, 1992) and firms employing less than 10 people increased their share of total employment from 19% in 1979 to 28% in 1991 (DTI, 1994). Over the period 1979–1988, the most dramatic growth in the numbers of businesses was in service sectors like finance and other business services, which achieved a net increase of over 60%, but the production industries also experienced a substantial net increase of 24% (Westhead and Moyes, 1992, p. 22). One of the features of the decline of the manufacturing sector since the mid-1960s has been the increasing relative importance of SMEs; the share of total enterprises for SMEs employing less than 200 workers increased from 19% in the 1960s to 32% in 1990 (Storey, 1994).

It is against this background of an increase in both the absolute and relative importance of small businesses within several national economies that the evidence relating to their performance in rural areas compared with urban areas needs to be considered. The focus will firstly be on the evidence relating to new firm formation, and secondly on the evidence relating to the growth and survival of existing small firms.

New businesses

Various studies have shown a clear urban–rural gradient in relation to the number of new business start-ups in the United Kingdom. Mason has described the variations as a north–south divide superimposed on a more detailed urban–rural contrast (Mason, 1991). Analysis at the county scale shows that it is the rural and semirural counties, located in both central and peripheral regions, which have the highest propensities to generate new firms (Fig. 8.3). By contrast, the conurbations (apart from Greater London) and the urban-industrial counties have the lowest rates. A recent cross-national study has shown that in the United Kindom the highest birth rates occurred in those rural regions which were adjacent to major urban areas (Reynolds *et al.*, 1994), thereby implying it is rural areas that are the most accessible to large towns and cities which are the most conducive to entrepreneurial activity.

Evidence from other countries also indicates spatial unevenness in new firm formation rates in favour of rural areas. Keeble and Wever (1986) show high rates (though lower volumes given their smaller populations) of new firm formation in a variety of previously unindustrialized rural regions, compared with low rates in urban, industrialized regions. In France the highest new firm formation rates are found in the rural regions of southern France (the Midi), there being a negative correlation ($r = -0.78$) between the proportion of the working population engaged in manufacturing and the rate of new manufacturing firm formation (Aydalot, 1986, p. 113).

Various writers have also identified a marked increase in the numbers of small manufacturing businesses in the rural regions of southern Europe (e.g. Lewis and

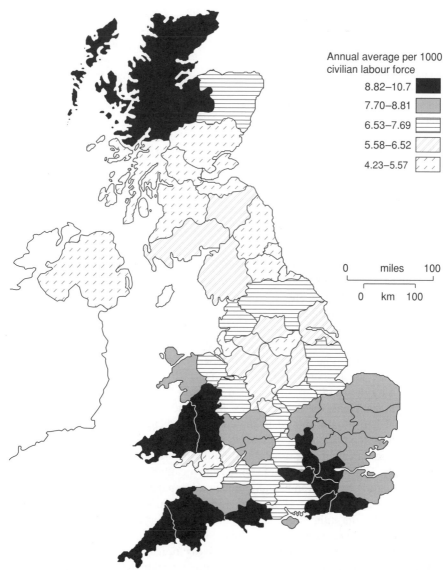

Figure 8.3 New firms: UK formation rates, 1980–1990 (Adapted from Keeble *et al.*, 1993, by permission of David Keeble)

Williams, 1986; Garofoli, 1992). In a study of rural small firms in the clothing sector of northern Greece, Simmons and Kalantaridis (1994) draw a clear connection between the restructuring of production associated with the drive for increased flexibility and the formation of new rural enterprises. They attribute much of the new business formation to the decisions of leading German clothing firms to move parts of their production to rural areas in southern Europe, where the wages are low. Thus the growth of small businesses in these rural areas is clearly linked to

Table 8.4   Migration origins of rural and urban business founders

	Company location		
Founder origin	Remote rural (%)	Accessible rural (%)	Urban (%)
Born in county	42.4	34.2	65.6
Moved to county before setting up firm	36.5	52.5	25.9
Moved to set up firm	21.1	13.3	8.6
Total	100.0	100.0	100.0

Source: Keeble and Tyler (1995).

the development of flexible production systems combined with the exploitation of new sources of cheap labour.

The evidence from several countries, therefore, would appear to indicate that rural areas are seen by potential entrepreneurs as being particularly favourable environments in which to start a new business. This raises the question of whether rural populations tend to be more 'entrepreneurial' than urban populations, or whether those seeking to establish a new business are moving from urban to rural areas in order to do so. At least as far as the United Kingdom is concerned, research has consistently shown that the majority of new business founders set up their businesses in the locality in which they are living (Mason, 1991). Although this is confirmed by the results of a major study entitled 'Business Success in the Countryside', it also points out that most new firm founders in rural areas were not born locally but had usually moved to the countryside before they set up their firms; whereas the proportion of entrepreneurs who were in-migrants was just over one-third in urban areas, it was around two-thirds in rural areas (Keeble *et al.*, 1992) (Table 8.4). Thus, this study claims to 'unequivocally demonstrate a direct connection between recent environmentally influenced population migration to England's rural areas and high rates of new enterprise formation there'. In a multivariate analysis of factors influencing new business formation rates at the county scale, Westhead and Moyes (1992) found population growth to be a key factor (achieving a correlation of 0.6 with the rate of new business formation, significant at the 0.001 level). Several researchers see a clear interrelationship between social changes taking place in rural areas and the in-migration of population on the one hand, mainly attracted by quality-of-life considerations, and economic development associated with new business formation on the other.

Existing firms

As well as displaying a better rate of new firm formation, rural areas also appear to have been doing well compared with urban areas in terms of the employment performance of existing small firms. In fact, one of the most striking findings of several recent studies in the United Kingdom has been that rural firms have shown a superior employment performance to comparable urban firms. A large national survey of over 2000 SMEs carried out by the University of Cambridge Small Business Research Centre (1992) showed that, over the period 1987–1990, rural firms had a median employment growth rate of 33.4% compared with 25% for firms in 'large towns' and 22.5% for those in conurbations. And over a similar duration, 1988–1991, Keeble *et al.* (1992) found on the basis of a matched-pair sample of both manufacturing and service firms that the fastest employment growth was in remote rural firms; their mean employment change was an additional 4.1 jobs per firm compared with 3.0 jobs per firm in accessible rural firms and a decrease of 1.7 jobs per firm in urban firms.

A longitudinal study of 306 SMEs in eight manufacturing sectors in three contrasting geographical environments during the 1980s has also shown that the rural firms had a much better employment performance (Smallbone *et al.*, 1993; North and Smallbone, 1995). Surviving rural firms achieved a net increase in employment of 51% over the period 1979–1990 compared with a 7% increase by similar firms located in London and 23% by their counterparts in outer metropolitan locations (in Hertfordshire and Essex). Despite differences in survey methodology and period covered, all three studies appear to agree that existing SMEs in rural areas are creating more jobs than SMEs in the same sectors in urban areas.

However, there is less agreement between researchers on the reasons for the better employment performance of rural SMEs. One view is that it reflects a greater interest in and ability to achieve business growth on the part of rural entrepreneurs. Another is that it has to do with the employment implications of the kinds of strategies which SMEs are adopting to achieve growth in different types of location as firms adapt to the constraints and opportunities which exist in their local operating environments. Each of these explanations will be considered in turn.

The theory of enterprising behaviour   Building upon the residential preference hypothesis for the urban–rural shift, Keeble and Tyler (1995) have recently advanced a 'theory of enterprising behaviour' to explain the superior employment performance of rural SMEs compared to their urban counterparts. Their theory is based upon two propositions. The first is that rural environments tend to attract people who are likely to be able to demonstrate a high level of enterpreneurial ability (the implication being that urban locations are less able to do this). This might seem surprising given the often expressed view that there are likely to be more 'lifestyle' businesses in rural areas. In a survey of the owner managers of new businesses in rural areas, Townroe and Mallalieu found that a significant proportion of entrepreneurs placed a higher priority on optimizing the benefits of being able to live in a rural environment than upon aiming for the growth and profitability of their business (Townroe and Mallalieu, 1993). The second proposition is that rural areas have an environment (in terms of economic, physical and institutional characteristics) which is particularly conducive to entrepreneurial behaviour. Thus, it is not just the residential environment of rural areas which is thought to be better, but the business environment as well.

These two propositions are thought by Keeble and Tyler to apply particularly to accessible rural areas rather than remote rural areas. This may account for the low

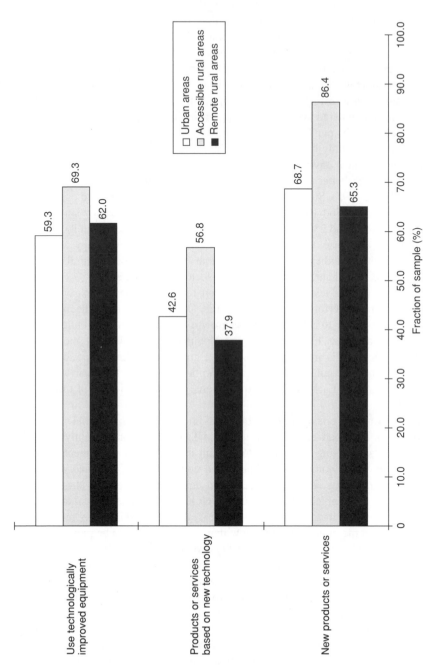

Figure 8.4 Urban–rural differences in innovation rates (Adapted from Keeble and Tyler, 1995, by permission of David Keeble)

incidence of 'lifestyle' businesses in their sample of firms drawn from what they defined as accessible rural areas (i.e. settlements of less than 10 000 population in the counties of Cheshire, Derbyshire, Northamptonshire and Wiltshire) as against the sample from remote rural areas (i.e. similar-sized settlements in several northern, eastern and south-western counties). A wide range of sectors were included in the study, including all manufacturing sectors and a range of professional, technical and business sectors.

Based on a sample of 1022 businesses, Keeble *et al.* (1992) have shown that firms in accessible rural areas had significantly higher ratings on a series of indicators measuring innovation, new products and technological expertise (Fig. 8.4), leading to the conclusion that 'accessible rural firms are more dynamic, innovative and technologically focused than their counterparts in either urban or remote rural locations' (Keeble and Tyler, 1995, p. 989). Accessible rural firms were also found to be more export oriented. According to the theory, the superior employment performance of rural SMEs therefore reflects the better quality of business owners and managers found in accessible rural areas, hence the more dynamic nature of the businesses themselves.

Urban–rural differences in small business growth strategies   A rather different slant on urban–rural differences in the employment performance of manufacturing SMEs results from the work of North and Smallbone (1995, 1996). This work compares SME performance during the 1980s in remote rural locations in northern England with the performance of similar firms in London and the outer metropolitan areas (OMAs) of Hertfordshire and Essex. Unlike the study by Keeble and Tyler, it does not include SMEs in more accessible rural areas; this may account for some of the differences in the results obtained. The main finding is that the urban–rural differences apparent in the employment performance of similar SMEs during the 1980s are not evident when growth is measured in output terms (Fig. 8.5, overleaf).

Not only did 39% of both the rural and the OMA samples more than double their sales turnover in real terms from 1979 to 1990, so did 35% of the London sample. And the proportion of the rural sample that remained stable or declined in terms of sales turnover (42%) was very similar to the proportion for the London sample (45%) and the OMA sample (48%). This similarity in the actual growth performance of urban and rural firms reflected very similar management attitudes to growth in these different geographical environments; 40% of rural firms had a clear growth objective during the 1980s compared with 41% of the London firms and 41% of OMA firms.

Contrary to the implications behind the results generated by those researchers who rely solely on employment to measure growth, these findings indicate there is a relatively small difference between rural and urban areas in the propensity for SMEs to grow in output terms. Instead, the difference relates to the employment implications of achieving growth; the growth of rural SMEs is more likely than the growth of urban SMEs to lead directly to employment generation. In other words, the growth of rural SMEs tends to be more labour-intensive than the growth of SMEs in urban and outer metropolitan locations.

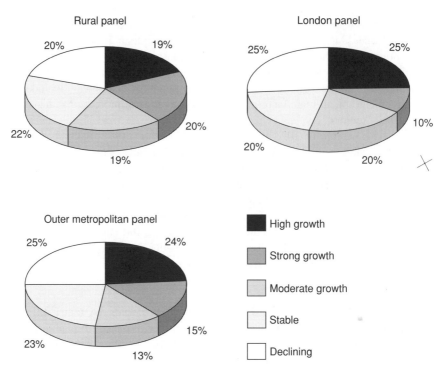

Figure 8.5   Distribution of SMEs by performance group in different types of location (Adapted from North and Smallbone, 1995)

Why should the growth of manufacturing SMEs in remote rural locations be more labour-intensive than the growth of their urban counterparts? A detailed comparison of the means by which SMEs achieved growth in the different geographical environments identifies several reasons, all of which relate to the ways in which SMEs adapt to their local operating environments; for a more detailed discussion see North and Smallbone (1996).

Key aspects of the local business environment that may influence the kind of approach which SME managers adopt to achieve growth are the nature of the local labour market and the kind of relationship they have with it. Firms from the same industrial sector may adopt very different growth strategies in rural environments than in urban environments, partly because they experience different labour market conditions. An important influence on the willingness of SMEs in remote rural locations to expand their workforces in line with their sales growth is the relative cheapness of workers in rural labour markets compared with workers in urban labour markets.

Comparison between rural and urban SMEs in terms of the wage levels for different occupations shows that, in printing for example, the median wage for a machine minder in 1990 was £195 in remote rural firms in northern England compared with £300 in inner London. Similarly, in furniture, the median wage for a skilled machinist in the remote rural firms was 78% of the wage in inner London.

This evidence therefore supports the view that the lower marginal cost of employing an additional worker in rural areas encourages firms to develop in ways that use labour more intensively. There is also less incentive to invest in labour-saving equipment in these circumstances, and the evidence does show some tendency for rural SMEs to lag behind their urban counterparts in the adoption of new and advanced production technology.

As well as these cost advantages, various qualitative attributes of rural workers also contribute to a more labour-intensive form of expansion, such as their reliability, commitment to the firm and flexibility. In many local labour markets the relative absence of alternative sources of employment undoubtedly contributes to low levels of turnover within the workforce, and low levels of unionization encourage greater functional flexibility within the workplace. Looking at SMEs from a manager's viewpoint, rural workers tend to be seen as compliant and adaptable. Given this generally high level of satisfaction with the quality of rural workers, there has been less inducement for rural SMEs to develop in ways which minimize their demand for additional labour (e.g. by overtime working or by externalizing activities to other firms); this was not the case with many urban SMEs.

A further example of how small business development can be influenced by the characteristics of the rural operating environment concerns the degree to which firms externalize various activities to other businesses. Because they are isolated from the main centres of industrial activity, the owner-managers of rural SMEs may be less aware of the opportunities that exist for subcontracting activities to other firms; even if they are aware of such opportunities, the additional transport costs and time involved might prove to be a deterrent. And insofar as subcontracting is often used to cut costs, rural SMEs may be under less pressure to do so as they do not face the same intensity of competition as their urban counterparts. The evidence from comparing the development paths of a sample of urban and rural SMEs certainly supports these expectations (North and Smallbone, 1995). In 1990 those regularly using subcontractors amounted to 48% of London firms and 53% of firms in outer metropolitan locations, compared to 30% of rural firms; moreover, just 11% of rural firms which increased their subcontracting during the 1980s did so as part of a cost reduction strategy compared with 40% of London firms and 32% of OMA firms. Compared with their urban counterparts, rural firms therefore have a higher propensity to grow by using internal resources and capacity.

Some time has been spent considering the growth of small businesses in rural areas in the United Kingdom, but this is justified in view of the key role they have been playing in rural industrialization. Detailed research at the level of the firm also provides some useful insights into the nature of the urban–rural shift. Explanations for the better performance of rural SMEs compared with urban SMEs have tended to focus on various behavioural characteristics of small business entrepreneurs, both in terms of the quality of the entrepreneurs who have been attracted to rural environments (particularly the more accessible ones), and in terms of how they have adjusted their growth strategies to the constraints and opportunities afforded by the rural operating environment. The more behavioural explanations for the urban–rural shift would therefore appear to be particularly appropriate to small businesses.

Other explanations, such as those which focus on capital restructuring and the labour needs associated with different stages of the product cycle, may be more appropriate to understanding the investment decisions by large firms in rural areas.

BRANCH PLANT INVESTMENT IN RURAL AREAS

Branch plant investment has played a particularly significant role in the industrialization of rural areas within newly industrializing countries such as Ireland. Whereas in the 1950s and 1960s much of Ireland's manufacturing industry was concentrated in the eastern part of the country, a convergence process has been occurring at the regional scale over the past 30 years. The rate of growth of manufacturing employment has been greatest in the most rural regions. In 1961 the level of industrialization, as measured by the proportion of the workforce in manufacturing, was 4.9 times higher in the East than in the West, respectively the most industrialized and the least industrialized regions at the time; but by 1986 the ratio had fallen to 1.6 (O'Malley, 1994). Much of this convergence has been the result of the locational preference of many foreign-owned companies (and most notably American-owned firms) for the less industrialized regions, particularly the West, the North-West and the mid-West. Thus foreign-owned firms account for just over 50% of manufacturing employment within these regions, compared with 43% in the Irish economy as a whole. Whereas much of the original inward investment was in technologically mature and labour-intensive sectors like clothing, footwear and textiles, more recent investment has tended to be in high technology sectors such as electronics, pharmaceuticals and medical instruments.

Once inside the EU boundaries, proximity to the market is not a crucial factor for these foreign-owned firms, since they are focusing on the export markets of the member countries of the EU rather than the Irish market *per se*. The key reasons for setting up in Ireland, especially in the more rural western regions, appear to be the low labour costs by European standards and the opportunity to draw upon a workforce with a lack of industrial experience and no tradition of unionization (O'Malley, 1994). Studies in other national contexts have also highlighted the importance of worker inexperience in the location of branch factories. In a study of high technology branch plants in non-metropolitan North and South Carolina and Georgia, Johnson (1989, p. 42) found that 60% of the factory managers characterized their workforces as being mainly comprised of workers lacking previous industrial inexperience.

Despite this lack of previous experience, surveys of foreign firms in western Ireland have shown a high degree of satisfaction with labour relations, the cost and availability of labour, and the levels of productivity achieved (Breathnach, 1985). And through the Industrial Development Authority (IDA), the Irish government has been particularly proactive in enticing multinational companies to western Ireland using a generous package of financial incentives, including investment grants and tax exemption on profits arising from export sales as well as a major advance factory construction programme. During the 1970s in particular, Ireland was seen as a 'tax haven' by many internationally integrated firms seeking to benefit from transfer pricing.

In fact, the level of expenditure on regional incentives (on a per capita basis in assisted regions) in Ireland is second only to that of Italy within the EU (Dignan, 1995, p. 82). The Irish example of rural industrialization, based largely on the locational decisions of multinational companies, provides considerable support for those who explain rural industrialization in terms of internationally mobile capital seeking out new 'green' sources of labour. From this standpoint, the role of the state in the form of the IDA can be interpreted as assisting the processes of capital accumulation by adding to the cost advantages of these rural locations.

But other evidence relating to branch plants in rural economies does put into question some of the ideas which are associated with both the product life cyle and the capital restructuring interpretations. In a survey of externally controlled branch factories in the rural region of Devon and Cornwall, which focused on the degree of restructuring that had occurred during the 1980s, Potter concludes there is little support for the traditional view that branch plants tend to concentrate on mature products and technologies and tend to employ a disproportionate number of low skill, non-unionized and female workers (Potter, 1995). Although only 10% of plants within the region are externally controlled, they tend to be the largest employers, accounting for 61% of manufacturing employment (Potter, 1993).

In the case of greenfield investments, the main factors behind the decision to locate in this rural region were once again the combination of labour advantages (particularly the availability and low cost of semiskilled labour and its non-militant nature), together with government financial assistance and the availability of sites and premises. Once they were established, however, Potter found that the majority of these branch plants were actively engaged in new product and market development and there was little evidence of them concentrating on mature and standardized products. Similarly, the majority of the plants were found to have been active in making changes in their manufacturing processes, changes designed to increase production flexibility.

In terms of changes in labour force skills, Potter found an increase in the proportion of skilled and technical workers, together with increased skill levels within the semiskilled workforce associated with a commitment to achieving enhanced product quality. And contrary to the expectations arising from the capital restructuring hypothesis, male workers were found to account for two-thirds of the workforce on average and over half of the plants had union recognition. This study therefore provides some important countervailing evidence to the kinds of images of rural branch plants that arise from the product life cyle and capital restructuring hypotheses; at least in Devon and Cornwall, branch plants appear to be investing in new product, market and process development, engaging in upskilling rather than deskilling of their workforces, and making an important contribution to rural economic development. Despite the undoubted pull of low cost labour in the first place, branch factories may be able to contribute more to strengthening rural economies and improving jobs prospects than some theories might lead one to believe.

CONCLUSION

This chapter has presented considerable evidence to show there has been a substantial growth of manufacturing industry in rural areas over the past quarter-century. Set against the context of a deindustrialization of the older industrial economies of western Europe and the United States, there has been a shift in the location of manufacturing within these economies away from the conurbations and large cities, and towards smaller towns and rural areas. It tends to have been the rural areas located closest to the major urban areas which have benefited most from this trend, particularly in capturing the more innovative high technology industries, although many remoter rural areas have also experienced an increase in their relative shares of manufacturing employment. Nor is the urban–rural shift confined to the older industrial economies, as shown by the newly industrializing countries of southern Europe, where there is evidence of a decentralization of manufacturing towards rural areas. The urban–rural shift is therefore to be found in a wide range of economic and political situations.

This chapter has contended there is no single explanation for the urban–rural shift; although several different theories have been put forward, they reflect the variety of overlapping and sometimes interrelated processes which are at work. Some theories are clearly more appropriate to understanding what is happening in certain situations than in others. Thus interpretations which focus on the influence of the quality of the residential and business environments provide convincing explanations for the growth of small businesses in rural areas; whereas those which focus on the reserves of cheap labour, easily moulded to the needs of the employer, are more applicable to those situations where large national and multinational companies have set up branch factories in rural areas. Urban and rural differences in land cost and availability have also played their part, although the ideas associated with the constrained location hypothesis were arguably more plausible in the 1960s and 1970s, when a relatively higher level of manufacturing investment was still occurring than in more recent times.

Finally, there is the question of whether or not one can expect the industrialization of rural areas to continue into the twenty-first century. It might be tempting to argue that the various efforts by governments and other agencies to regenerate urban economies over the past couple of decades will lead to a reindustrialization of many cities in the developed economies, reducing the scale of the urban–rural shift if not completely reversing it. The extent to which this occurs is likely to depend on the degree to which urban areas have improved their ability to provide for the industrial space needs of modern businesses as well as their success in overcoming the negative externalities frequently associated with an urban location. However, there are other, arguably more convincing, reasons for expecting a continuation of rural industrialization. As in southern Europe, the trend towards more flexible production systems is leading to businesses in core regions externalizing their more labour-intensive activities to small businesses in peripheral rural areas. This trend is likely to increase as production systems become more internationalized, aided by developments such as the EU's Single Market.

Possibly the most compelling reason for expecting the industrialization of rural areas to continue into the twenty-first century has to do with advancements in information and communication technologies and their implications for the competitiveness of rural economies. In particular, developments in telematics (i.e. the fusing together of telecommunications and computer technology) are minimizing if not completely eliminating the effects of time and distance upon business operation, thereby opening up all kinds of new possibilities for businesses in what have tended to be regarded as marginal locations. The use of telematics will make it easier for these businesses to penetrate non-local markets, to participate in business networks, and to make use of product, market and other business databases. In other words, it has been argued that some of the main disadvantages associated with remote rural locations will be overcome, thereby increasing the competitiveness and growth potential of rural businesses (Grimes, 1992). Added to which, telematics is likely to make it possible for an increasing number of business people to live and work in a rural setting, thereby accentuating the importance of the quality of the residential environment on the location of new and small businesses, as well as upon the self-employed (i.e. the rural teleworkers).

Once again, however, the issues are more complex than at first they may seem. In theory telematics does open up these new opportunities for rural businesses, but it depends upon at least two key barriers being overcome. Firstly, there is the geographical distribution of investment in the fibre-optic networks themselves (the so-called information superhighway) and the risk that rural areas will be bypassed because of the much greater financial returns to be gained from concentrating investment close to the major population and business centres. Although rural businesses are able to gain access to various services (e.g. fax, eletronic mail, electronic data interchange) via the telephone system, there is concern they may be isolated from the latest developments in broadband cable networks.

The second barrier which has to be overcome is the relatively low rate of adoption of telematic services that are universally available and the underutilization of computers, revealed by various surveys of rural businesses (e.g. Clark *et al.*, 1995). This suggests that much greater investment in telematics awareness and skills training is going to be necessary if rural businesses are to stand a chance of grasping the opportunities which new technology brings. These two barriers to the diffusion of telematics in rural areas show that one cannot predict with certainty whether these latest technological changes will reinforce the process of rural industrialization, since much will depend upon the particular national context, including government policy towards the spatial diffusion of technology.

REFERENCES

Allen, J. and Massey, D. (1988) *The economy in question*. Sage, London.

Aydalot, P. (1986) The location of new firm creation: the French case. In Keeble, D. and Wever, E. (eds) *New firms and regional development in Europe*. Croom Helm, London.

Barkley, D. (1988) The decentralisation of high-technology manufacturing to nonmetropolitan areas. *Growth and Change*, **19**(1), 13–30.

Breathnach, P. (1985) Rural industrialisation in the west of Ireland. In Healey, M. and Ilbery, B. (eds) *The industrialisation of the countryside*. GeoBooks, Norwich.

Clark, D., Ilbery, B. and Berkeley, N. (1995) Telematics and rural businesses: an evaluation of uses, potentials and policy implications. *Regional Studies*, **29**(2), 171–80.

Dicken, P. (1992) *Global shift: the internationalisation of economic activity*, 2nd edn. Paul Chapman, London.

Dignan, T. (1995) Regional disparities and regional policy in the European Union. *Oxford Review of Economic Policy*, **11**(2), 64–95.

DTI (1994) *Small firms in Britain 1994*. HMSO, London.

Employment Department (1992) *Small firms in Britain report 1992*. HMSO, London.

Erickson, R. (1976) The filtering down process: industrial location in a non-metropolitan area. *Professional Geographer*, **26**, 254–60.

Estall, R. (1983) The decentralisation of manufacturing industry: recent American experience in perspective. *Geoforum*, **14**(2), 133–47.

Fothergill, S. and Gudgin, G. (1982) *Unequal growth: urban and regional employment change in the UK*. Heinemann, London.

Fothergill, S., Kitson, M. and Monk, S. (1985a) *Urban industrial change*. HMSO, London.

Fothergill, S., Gudgin, G., Kitson, M. and Monk, S. (1985b) Rural industrialisation: trends and causes. In Healey, M. and Ilbery, B. (eds) *The industrialisation of the countryside*. GeoBooks, Norwich.

Garofoli, G. (ed) (1992) *Endogenous development in southern Europe*. Avebury, Aldershot.

Gould, A. and Keeble, D. (1984) New firms and rural industrialisation in East Anglia. *Regional Studies*, **18**(3), 189–202.

Grimes, S. (1992) Exploiting information and communication technologies for rural development. *Journal of Rural Studies*, **8**, 269–78.

Gudgin, G. (1995) Regional problems and policy in the UK. *Oxford Review of Economic Policy*, **11**(2), 18–63.

Hodge, I. and Monk, S. (1987) Manufacturing employment change within rural areas. *Journal of Rural Studies*, **3**(1), 65–69.

Johnson, M. (1989) Industrial transition and the location of high-technology branch plants in the non-metropolitan southeast. *Economic Geography*, **65**(1), 33–47.

Keeble, D. (1980) Industrial decline, regional policy and the urban–rural manufacturing shift in the United Kingdom. *Environment and Planning A*, **12**(8), 945–62.

Keeble, D. (1992) High technology industry and the restructuring of the UK space economy. In Townroe, P. and Martin, R. (eds) *Regional development in the 1990s*. Jessica Kingsley, London.

Keeble, D. (1993) Small firm creation, innovation and growth and the urban–rural shift. In Curran, J. and Storey, D. (eds) *Small firms in urban and rural locations*. Routledge, London.

Keeble, D. and Tyler, P. (1995) Enterprising behaviour and the urban–rural shift. *Urban Studies*, **32**(6), 975–97.

Keeble, D. and Wever, E. (1986) *New firms and regional development in Europe*. Croom Helm, London.

Keeble, D., Owens, P. and Thompson, C. (1983) The urban–rural manufacturing shift in the European Community. *Urban Studies*, **20**, 405–18.

Keeble, D., Tyler, P., Broom, G. and Lewis, J. (1992) *Business success in the countryside: the performance of rural enterprise*. HMSO, London.

Keeble, D., Walker, S. and Robson, M. (1993) *New firm formation and small business growth in the United Kingdom: spatial and temporal variations and determinants*. Employment Department Research Series 15.

Lewis, J. and Williams, A. (1986) Factories, farms and families – the impacts of industrial growth in rural central Portugal. *Sociologia Ruralis*, **26**(3/4), 320–44.

Lonsdale, R. (1985) Industrialisation of the countryside: the case of the United States. In Healey, M. and Ilbery, B. (eds) *The industrialisation of the countryside*. GeoBooks, Norwich.

Lonsdale, R. and Seyler, H. (eds) (1979) *Non-metropolitan industrialisation*. Halsted Press, New York.

Markusen, A. (1985) *Profit cycles, oligopoly, and regional development*. MIT Press, Cambridge MA.

Mason, C. (1991) Spatial variations in enterprise: the geography of new firm formation. In Burrows, R. (ed) *Deciphering the enterprise culture*. Routledge, London.

Massey, D. (1984) *Spatial divisions of labour: social structures and the geography of production*. Macmillan, London.

North, D. and Smallbone, D. (1993) *Small business in rural areas*, Rural Development Commission Strategy Review Topic Paper 2. RDC, London.

North, D. and Smallbone, D. (1995) The employment generation potential of mature SMEs in different geographical environments. *Urban Studies*, **32**(9), 1517–34.

North, D. and Smallbone, D. (1996) Small business development in remote rural areas: the example of mature manufacturing firms in northern England. *Journal of Rural Studies*, **12**(2), 151–67.

O'Malley, E. (1994) The impact of transnational corporations in the Republic of Ireland. In Dicken, P. and Quevit, M. (eds) *Transnational corporations and European regional restructuring*. Netherlands Geographical Studies 181, Utrecht.

Piore, M. and Sabel, C. (1984) *The second industrial divide: prospects for prosperity*. Basic Books, New York.

Potter, J. (1993) External manufacturing investment in a peripheral rural region: the case of Devon and Cornwall. *Regional Studies*, **27**(3), 193–206.

Potter, J. (1995) Branch plant economies and flexible specialisation: evidence from Devon and Cornwall. *Tijdschrift voor Economische en Sociale Geografie*, **86**(2), 162–76.

Reynolds, P., Storey, D. and Westhead, P. (1994) Cross-national comparisons of the variation of new firm formation rates. *Regional Studies*, **28**, 443–56.

Rural Development Commission (1993) *Rural development strategy for the 1990s*. RDC, London.

Scott, A. and Storper, M. (1988) The geographical foundations and social regulation of flexible production systems. In Wolch, J. and Dear, M. (eds) *Territory and social reproduction*. Allen and Unwin, London.

Shutt J., and Whittington, R. (1987) Fragmentation strategies and the rise of small units: cases from the North West. *Regional Studies*, **21**(1), 13–23.

Simmons, C. and Kalantaridis, C. (1994) Flexible specialisation in the southern European periphery: the growth of garment manufacturing in Peonia County, Greece. *Comparative Studies in Society and History*, **36**(4), 649–75.

Smallbone, D., North, D. and Leigh, R. (1993) The growth and survival of mature manufacturing small and medium-sized enterprises in the 1980s: an urban–rural comparison. In Curran, J. and Storey, D. (eds) *Small firms in urban and rural locations*. Routledge, London.

Storey, D. (1994) *Understanding small business*. Routledge, London.

Storper, M. and Walker, R. (1984) The spatial division of labour: labour and the location of industries. In Sawyers, L. and Tabb, W. (eds) *Sunbelt/snowbelt: urban development and regional restructuring*. Oxford University Press, New York.

Thrift, N. (1987) Manufacturing rural geography? *Journal of Rural Studies*, **3**(1), 77–81.

Townroe, P. and Mallalieu, K. (1993) Founding a new business in the countryside. In Curran, J. and Storey, D. (eds) *Small firms in urban and rural locations*. Routledge, London.

Townsend, A. (1993) The urban–rural cycle in the Thatcher growth years. *Transactions of the Institute of British Geographers*, **NS18**, 207–21.

Tyler, P., Moore, B. and Rhodes, J. (1988) Geographical variations in industrial costs. *Scottish Journal of Political Economy*, **35**, 22–50.

University of Cambridge Small Business Research Centre (1992) *The state of British enterprise: growth, innovation and competitive advantage in small and medium-sized firms*. Small Business Research Centre, University of Cambridge.

Westhead, P. and Moyes, T. (1992) Reflections on Thatcher's Britain: evidence from new production firm registrations 1980–88. *Entrepreneurship and Regional Development*, **4**, 21–56.

CHAPTER 9

POLICIES AND PLANNING MECHANISMS:
Managing change in rural areas
Andrew Gilg

INTRODUCTION

Rural change does not take place in a vacuum but in an atmosphere richly influenced by political systems. In developed market economies these political systems have accepted that the free market will continue to be the main vehicle for development, but that planners will need to intervene by guiding and shaping the free enterprise system in order to make it both more efficient and less harmful to the environment. In recent years this vision of rural planning policy has focused around the concept of 'managing change', and managing change in rural areas has become a key aim of planners at all levels of government.

Attempts to manage change have been made across all the topics covered by this book, notably in agriculture, forestry, and nature conservation. Acordingly, these attempts are not considered in this chapter, but the material presented here should be read bearing in mind the effect that planning policies for other rural sectors have had on managing or preventing change to built environments in rural areas.

Managing change rests on two principal assumptions: (1) the managers have a vision of what is to be achieved; (2) they have the instruments to achieve it. Managing change thus contains two vital ingredients: (1) a set of *policies* derived from the vision; (2) a set of *implementation* tools for achieving policy aims. These two ingredients should be seen as two parts of the wider overall policy process by which rural environments are regulated (Flynn and Murdoch, 1995).

The rest of this chapter is thus divided into two main sections: a section on policy examines a model of the process from first principles; and a section on implementation looks at two case studies of contrasting market economies, because planning systems are rooted in the culture which gives them life. The first is the United Kingdom, a centralized state with strong but basically discretionary controls over development, the second is the United States, a federal state with weak controls over development but pragmatic and innovative methods of managing change. The United States can also broadly be used as a metaphor for the zoning systems used by most other industrialized countries.

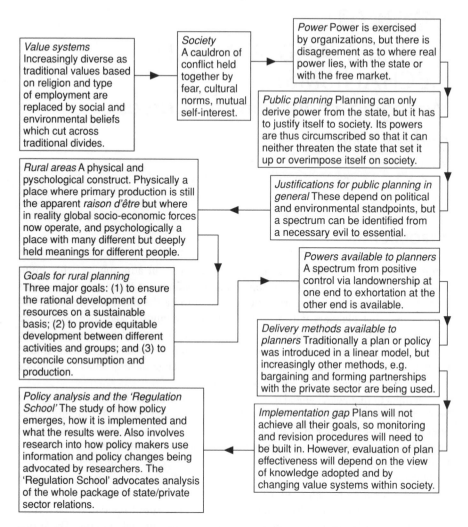

Figure 9.1   Simplified model of the relationships between society, power and planning policies: the arrows show the flow used in this chapter; they have no other function (Adapted from Gilg, 1996)

POLICY

Politics is the art of the possible, so policy derived by politicians must be set in the context of the society within which it operates. It is therefore appropriate to begin with a simplified model of planning policy (Fig. 9.1).

Value systems   Value systems are the core of any society and they determine the vision behind policy. Traditionally, the value systems of Western liberal democracies were based on Judaeo-Christianity and the work ethic. Most of the institutions in these countries are therefore based on a set of values which reward economic success. In recent years, however, these 'absolute' values, notably those of Roman Catholicism, have been queried by a set of values based on 'relativism' and permissiveness. In addition, societies have become more multicultural. A growing gap has thus occurred between slow-moving institutions and an increasingly diverse and fluid set of values. Ironically, this growing diversity increases the need for rural policies, since one of their roles is to reconcile conflict between competing land uses and thus prevent society from becoming ever more fissile.

Society Society can be defined as the matrix of value systems which provide enough common factors to hold groups of people together in peace and some sort of harmony. Society thus implies a state of semi-equilibrium where people have foregone individual freedoms and desires, and crucially have surrendered many powers over their own destiny to organizations which can impose norms and regulations on them. Unfortunately, 'modernity', the creation of a technological, secular and segregated world, has weakened social ties and alienated many groups who are only kept in society by the fear of sanctions. The concept of society as a cauldron of conflict held together by sanctions as much as by cultural norms and self-interest is thus crucial to understanding rural policy. Such policies have to assume some common value systems around which to focus attention, and the degree of compliance or deviation that can be expected, depending on how far these policies accord with all or only a few sections of society.

Power Power can be seen as imposed from above by powerful groups or as a set of freedoms willingly surrendered in the common good to the state and its agency, the government, which governs by consent. The United Kingdom and the United States offer contrasting views here (Cullingworth, 1993). The United Kingdom began as a monarchy, from which concessions have been slowly but surely wrung ever since Magna Carta in the thirteenth century, in a series of ad hoc changes. The British system of policy is thus based on precedent and incrementalism, and there is no constitution. In the twentieth century, Parliament has become apparently pre-eminent, but in reality it shares power with the corporate grouping of public institutions, industrial firms and financial services that make up the British state. The individual elector, either through the ballot box or through presure groups, has little say except cosmetically.

But in the United States, the government was set up after two sets of revolutionary events: first, the emigration of people from the British Isles to be free from the restrictive power of the monarchy and imposed religions; and second, the war of independence from Britain, which resulted in the first modern constitution, based crucially on the freedom of the individual and the right to use private property without hindrance (Jackson, 1986). People in the United States thus resent political policies as an intrusion into their liberty. In addition, the way Congress works makes it difficult to get controversial legislation passed. First, the president may well be leading a minority party, and second, Washington has become dominated by powerful lobby groups in the postwar years, and any public institution is subject to the pressure of a thousand voices (Christensen, 1993). Americans are also unconvinced about the need for environmental policies because the American Dream of 'making a million' and moving on across the unlimited continent remains deep in their psyche (Held and Visser, 1982). In the United States, power is seen to be conceded as a last resort, and the vagaries of free enterprise are tolerated because its benefits are perceived to outweigh its disadvantages (Lassey, 1977).

However, in an increasingly global world, national governments have had to concede some power to global and international organizations. Most notably the 1992 United Nations Conference on the Environment and Development led to nation states agreeing to implement 'sustainable' policies in their own countries. In addition, other global agreements have placed restrictions on energy use, with enormous implications for the extensive growth of settlements and other land uses, notably in the United States, dependent on automobile use. Internationally, the European Union has several environmental programmes which have impinged on UK rural planning policy, most notably the concepts and practices of environmental assessment. However, both global and international agreements have also emphasized that local action is crucial, as exemplified by the European Union's concept of 'subsidiarity', which advocates that decisions are taken at the most local level possible, and in the aphorism: Think globally, act locally.

Policies, therefore, should ideally be developed by a set of interconnected organizations at four levels – global, international, national and local – with decisions flowing up and down easily. However, in the real world, power tends to be exercised in a top-down fashion, notably in the United Kingdom, which is a very centralized country. In contrast, the United States is a federal country, where the centre has certain powers, but much power is devolved to the states, individual counties, and local administrations or municipalities. There are benefits and drawbacks to both systems. For example, the United Kingdom has the merit of a national system where both developers and the public know where they stand, but where individual local authorities have considerable discretion in decision making within the context of national laws. In the United States, however, there is much discretion and flexibility, but by the same token no uniformity and the possibility of playing off one area against another, leading to uncoordinated and ultimately inefficient patterns of land use.

At both national and local levels, power is theoretically exercised by politicians. In reality, at the national level, politicians act within the constraints imposed by the need to keep free enterprise prosperous in an increasingly competitive world, and are thus unwilling to impose too many land-use controls. But at the local level, planners and politicians are frequently under great pressure to restrict development in order to protect the rural environments of existing residents. So they have to modify policies that may be of benefit to the wider community in order to pacify the demands of influential sections of their electorate.

Power is also exercised differentially depending on political ideologies. Left-wing and centre parties traditionally seek solutions via public policy, especially those based on regulation and control. But right-wing parties, which have dominated most market economies in recent years, dislike public policy, and when they deem it necessary, they favour incentive policies based on financial or fiscal measures. Traditionally, no ideology holds sway for much more than a decade, so public policy tends to be marked by a series of shifts across the political spectrum.

Power is thus an elusive, but very contested concept. Its exercise is difficult and has to concede many compromises to different interest groups, albeit within the context of the free market. Neoclassical (traditional) economics takes it for granted that the laws of supply and demand will provide the vehicle for compromise. In recent years this view has been challenged by the regulation school, which argues that the expanded social reproduction of capitalism is never guaranteed and continually has to be secured through a range of social norms, mechanisms and institutions which help to stabilize the system's inherent contradiction temporarily around a particular regime of accumulation (Goodwin *et al.*, 1995). From the evidence of this century, however, it can be concluded that public policy evolves in response to a random sequence of events, including major world trends, scientific and technological advances, charismatic leaders and pragmatism, rather than the allegedly conspiratorial powers of capitalism. In summary, disjointed and accretionary incrementalism based on a variation of Samuel Butler's famous aphorism that life (policy development) is the art of drawing sufficient conclusions from insufficient premises (data), seems to have been the order of the day.

Public planning Public planning axiomatically derives its power from the state; it therefore has to tread a difficult path between biting the hand that gave it life and being so toothless that it might as well not exist. This is because most demands for planning have come from third parties who wish to control the actions of landowners and developers. The only way they can exercise control is via some sort of planning system. Although the main dilemma faced by most planning systems is the balance between the interests of economic development and environmental conservation, they are also faced with trying to impose the concepts of a common good on a diverse and pluralist society. Indeed the post-modern school of thought has criticized the concept of 'utilitarianism' so widely employed by planners as being no longer viable. Thus planners need once again to justify themselves.

Justifications for public planning in general Justification sprang initially from the negative externalities of unplanned industrial growth, which led to the disease, squalor and congestion of nineteenth-century cities. Since then planners have attempted to expand their role by advancing a range of arguments across a fourfold spectrum, ranging from planning as a necessary evil to planning as an essential activity (O'Riordan, 1985).

First, the view of planning as a *necessary evil* is the one advanced by right-wingers and is centred around negative controls preventing bad things. Key examples are stopping or reducing environmental damage; preventing chaos; preserving property rights and values; protecting the poor and weak from exploitation; and preventing unattractive environments or inefficient settlement patterns and related land uses being created.

Second, the view that planning should be *reactive* is the one advanced by the left wing of right-wing parties and the right wing of left-wing parties. The arguments are focused around making the market more efficient or correcting market failures

(Lassey, 1977). More left-wing arguments centre on the market's failure to take environmental goods and losses (pollution) into account. Typical arguments include seeking consensus; resolving existing conflicts; preventing possible conflicts; balancing economic growth with the preservation of the environment; balancing efficiency with equity; managing change; coordinating development; and seeking the greatest good for the greatest number (utilitarianism). Key principles which follow from this reactive aproach are the precautionary principle; the polluter pays principle; the concept of the best practicable environment option; and sustainability. The arguments are thus pragmatic ones about making capitalism better.

Third, the view that planning should be *proactive* is the one advanced by socialists. Typical arguments here include putting the community in control; making things happen that wouldn't otherwise; and making things better (utopianism), one of the arguments of the founders of planning. Fourth, the view that planning is *essential* is the one advanced by communists and/or deep green ecologists. Both agree that capitalism and uneven development are the root cause of both social and environmental problems. Communists advocate a massive transfer of productive resources to the state, hence the transformation of planning from a third-party indirect activity to a direct state activity with all the apparatus this implies. In contrast, deep green ecologists argue that draconian controls are needed if eco-catastrophe is to be avoided.

Each element of this spectrum is not mutually exclusive and in practice a combination of arguments is advanced, depending on the circumstances. Rural areas in particular have produced their own set of arguments for public planning.

Rural areas   The concept of *rural areas* is both elusive and complex, and it has been changing over recent years. Cloke *et al.* (1994) have identified four phases. The first phase equated rurality with particular spaces and functions, notably extensive land uses and low population densities. The second phase replaced these static concepts with the more dynamic concept of political economy in which rural life was seen as a power struggle between class interests (Marsden *et al.*, 1993). The third phase centred around post-modern notions of rurality as a social construct reflecting a world of social, moral and cultural values. The fourth phase has used post-structuralist deconstruction in an attempt to understand the symbolic meanings of rurality for people. Ironically, this has shown that many people use the functional definitions employed in the first phase (Halfacree, 1995). Indeed, each phase should not be seen as mutually exclusive but as an alternative viewpoint with its own merits. In light of this and the plurality of views held by individuals, Jones (1995) has argued that rural areas should now be seen as a melee of conflicting interpretations. The increasingly dynamic, global, and culturally confused nature of rural areas thus poses severe problems when setting goals for rural planning.

Goals for rural planning   Goals are thus almost limitless. But if various attempts over the years are agregated (Lassey, 1977; Hill and Young, 1991; McDonald, 1989; Gilg, 1991) then three major aims stand out: (1) to ensure the rational development of resources on a sustainable basis; (2) to provide equitable development between different activities and groups within society; (3) to reconcile consumption and production. In terms of planning the built environment of rural areas, these major aims translate into protecting farmland, conserving the environment, and preventing wasteful land-use patterns based on the separation of home, work and leisure activities.

Powers available to planners   In common with the justifications for public planning, powers can be placed in a spectrum based on political ideologies. Working independently, Gilg (1991) and Selman (1988) have both produced spectra which can be modified as shown below:

1. Voluntary methods based on exhortation, advice and examples of good practice.
2. Financial incentives to encourage desirable land uses.
3. Financial disincentives to discourage undesirable uses.
4. Regulatory controls on developing land or land uses.
5. Public ownership or land management via long-term leases.

Traditionally, planners have moved along the spectra, threatening the next type of power unless improvements are made. Most planning systems employ all elements of the spectrum in a rich variety of combinations. In addition, planners can employ three other options: (1) they can set up some sort of administrative organization; (2) they can set up special areas; (3) they can explore innovative systems, such as putting the right to develop land up for public auction or competition.

Delivery methods available to planners   Delivery methods have traditionally focused on the classical process of plan making followed by plan implementation, and plan revision (Fig. 9.2, overleaf). This process has two distinct operations: setting out policy, then implementing it by the control of development in a process known as the rational comprehensive linear systems approach. Since the heyday of this approach in the 1960s it has been attacked as idealistic, inoperable and attempting to impose norms and standards that are not generally agreed. In particular, the political economy school has attacked these planning systems as merely legitimizing capital accumulation (Short *et al*., 1986) and the post-modern school has criticized the 'master meta-narrative' approach (Beauregard, 1991) adopted by most planning documents and their extravagant claims to speak on behalf of 'the public' (Tett and Wolf, 1991). The search for alternatives has been extensive, notably in the academic literature (Gilg, 1991), but real-world planners have adapted the rational linear systems approach in two ways. First, they have substituted the idea of the utopian end-state plan, with the concept of the plan as a framework and starting-point for negotiation. Second, they have modified their role as experts who hand down scientifically perfect, value-free judgements and substituted the role of arbitrator between all interested parties, seeking consensus through a process of bargaining and negotiation. The essential feature of the planning delivery system is thus to manage change by muddling through; this is the concept of 'disjointed incrementalism' developed by Lindblom (1977).

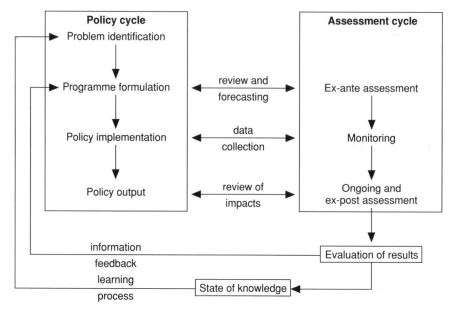

Figure 9.2   Links between policy cycle and assessment cycle (Reprinted from Schrader, 1994, by permission of Elsevier Science Ltd.)

Implementation gap   An *implementation gap* (Cloke, 1987) appears to exist in all planning systems between policy and implementation. Indeed, the possibility of plan failure was explicit in the traditional model. What has altered is the greater diversity in the ways of looking at planning, and of opinions about what planning should be attempting to do. In addition, politicians have increasingly adopted an accountancy or 'bottom line' approach to evaluating the effectiveness of public policies, like planning, and are looking at performance indicators or targets by which to measure the implementation gap. Planning policies are now increasingly couched in terms that allow performance to be measured.

Policy analysis and the 'Regulation School'   Policy analysis is a growing field which predated the modern obsession with the 'bottom line' approach, but has fed on it. According to Ham and Hill (1984), it can be divided into two activities. First come studies of the policy process. In more detail they involve a study of the various influences on policy formulation, a study of the outputs of policy (e.g. number of houses built), and a study of impacts which may be descriptive and/or prescriptive. Second come studies for the policy process. These involve researchers working with policy makers to see how they might make better use of information in policy formulation. The 'Regulation School of Thought' (Goodwin *et al.*, 1995) goes further and insists that one looks at the total package of relations and arrangements which contribute to the stabilization of output growth and aggregate distribution of income and consumption. This derives from work by Jessop (1990) who, in his development of regulation theory, has identified the policy process as a

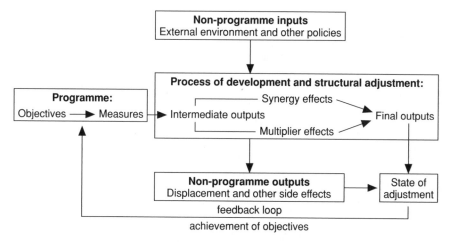

Figure 9.3 Impact assessment and causality (Reprinted from Schrader, 1994, by permission of Elsevier Science Ltd.)

Fordist mode of regulation in which the complex of regulatory mechanisms, procedures and institutions built around collective norms, rules, habits, laws and customs, sustains and supports the macrolevel regime of accumulation. However, in a critique of this approach, Clark (1992) has called for a more pragmatic concept of 'real' regulation which places as much emphasis on social forces in the policy process as economic forces.

Although policy analysis and regulation theory provide useful organizing concepts, which can be used as shown in Fig. 9.3, there are several difficulties with them, notably in linking cause and effect (Pearce, 1992). First, there is no general agreement about how to define policy. Second, it is hard to identify particular occasions when policy is made. Third, policies may be set out in a variety of forms, such as policy documents and legal decisions. Fourth, policies may not be prioritized and contradictory policies may exist not only between policy areas but even within policy areas. Fifth, policies are subject to change, leading to a time lag and overlapping of often contradictory policies. Sixth, decision makers will be affected by the policy and will alter their behaviour by either 'displacing' it somewhere else or not doing it at all, in what may be called the 'AIDS' and 'birth control' effects (Gilg and Kelly, 1996). Seventh, policies may be ignored by decision makers either through ignorance or wilfully; and eighth, a series of ad hoc decisions can be post hoc rationalized and developed into policy in order to legitimize past decisions.

Nonetheless, Tarrant (1992) has set out three approaches to asessing policy effectiveness on the ground: (1) comparing two similar areas, but areas subject to different policies; (2) by simulating what might have happened without the policy, or with a different policy, then comparing this with what actually happened; (3) by examining what happened with the policy predictions. Methodologically, this can be done by statistical manipulation of data, by interviewing people, especially in a behavioural, humanistic or post-modern framework, or by setting the policy in a

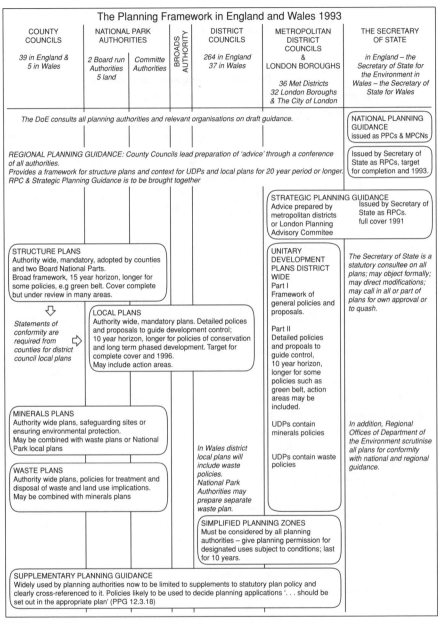

Figure 9.4   The planning framework in England and Wales, 1993 (Reprinted from Cullingworth and Nadin, 1994, by permission of Routledge)

political economy or structural framework and studying the interactions that take place (Cloke *et al.*, 1991).

Summary   This section has set out the context of public policy for rural areas, notably managing change to a built environment. It has described the powers and delivery systems that may be used, and the problems of evaluating the effects of policy. It is now time to move from a general consideration of policy to some case studies of implementation drawn from the United Kingdom and the United States.

IMPLEMENTATION: UNITED KINGDOM

Three core concepts

The planning system in the United Kingdom has rested on three core concepts since it was inaugurated in 1947. The first core concept is that planners try to guide and shape development by positive guidance in plans and policy documents, backed up by the power to refuse, impose conditions or unconditionally allow proposals for development. Since 1947 the right of private interests to develop land as they see fit has been conceded in the public interest. The second concept is that the refusal or imposition of conditions does not lead to the payment of compensation, although a one-off fund was set aside to pay compensation in the years immediately following 1947 for those people who had bought land before 1947 in the expectation of developing it. The third concept is that planning as conceptualized above is accepted by all political parties; even the Conservative Party in their most radical period in the early 1980s argued:

> The planning system balances the protection of the natural and the built environment with the pressures of economic and social change. The need for the planning system is unquestioned and its workings have brought great and lasting benefits. (DoE, 1980)

The planning framework

The three core concepts are operated within the planning framework shown in Fig. 9.4. The key features of this framework are that the Department of the Environment (DoE) sets the national standard via primary legislation and its detailed interpretation via statutory instruments and circulars, and by adjudicating over appeals either in-house or normally through its executive agency, the Planning Inspectorate. Detailed land-use policies are set by Local Plans, produced by district councils working within the policy context set by Structure and Regional Plans. In particular, they provide a guide to how many houses could be built and where they could be built. These plans provide the basis on which to make decisions about applications to develop land. This process of development control is the heart of the British planning system and its unique aspect since, although policies provide the basis for decisions, each case is still treated on its merits. Thus the process is not as mechanistic as most planning systems which are based on zoning. Development control in Britain can and does provide contrary decisions to policy.

In rural areas, the key aims of development control have traditionally been to protect farmland, to preserve attractive landscapes and to concentrate development into a few key settlements which can be serviced most cost-effectively (Cloke and Shaw, 1983). However, in recent years these aims have gradually changed as agriculture has declined as an employer, with a consequent need to find alternative employment.

Nonetheless, the central ethos is still one of protecting the countryside and this has recently been given a powerful boost by policies against out-of-town developments based on the motor car, inspired by the 1992 Earth Summit in Rio de Janeiro. The need to conserve the countryside in order to produce food, which stemmed from the food shortages of World War II and the immediate postwar period, has been replaced with the need to conserve the planet from exhaust emissions and to avoid the horrors of sprawl, as experienced in North America.

Managing or preventing change

The first, but disputed, impact has been that the rate of land loss to urban growth experienced in the 1930s has markedly slowed down. According to the DoE (1992), the rate had fallen from 25 000 hectares per year in the 1930s to only 5000 hectares per year by the late 1980s. However, this is disputed by Sinclair (1992), who has used a combination of data to calculate that the average loss between 1945 and 1990 was 15 700 hectares, with a high of 21 000 hectares in the 1960s and a low of 11 000 hectares in the late 1980s.

The process has, however, been very uneven and the main resistance to development in the countryside has come increasingly from communities in the green belt and the countryside immediately surrounding the major cities, notably London. Short *et al.* (1986), in a study to the west of London, have demonstrated how local people have attempted to frustrate development, only to be overruled by central government, which has insisted on economic growth in one of the most dynamic areas of the United Kingdom. Murdoch and Marsden (1994), in a study to the north-west of London, have shown how resistance to urban growth has strengthened and how middle-class values have been used to keep the countryside undeveloped. Nonetheless, in spite of the fact that megalopolis has been denied in the United Kingdom (Hall, 1974), the countryside has experienced considerable population growth in certain selected settlements beyond the green belts (Champion, 1993), and a form of 'spread city' has developed in a spider's web of connectivity across a wide arc of south-east England encompassed by Southampton, Swindon, Cambridge and Norwich (Hall, 1988).

This growth, however, has not taken place in the open countryside; it has been confined to a few settlements. These settlements have not always been chosen by planners (Blacksell and Gilg, 1982; Cloke, 1979), and about half of all villages have seen populations decline (Weekly, 1988). These restrictive policies have had regressive social effects, with high house prices leading to a rapid gentrification of the population in some areas (Harper, 1987) and the rise of a new 'service class' of rural inhabitants elsewhere (Cloke and Thrift, 1987) which has reinforced traditional attitudes against development in rural areas.

Protected landscapes, ironically, are subject to even more intense pressure for development which arises from their designation (Brotherton, 1982). However, the 20% of England that is protected as either national park or area of outstanding natural beauty has resisted development pressure, except from some government departments, and their built environments have been preserved from development (Anderson, 1981; Blacksell and Gilg, 1982).

In summary, Gilg (1996) has argued that the planning system has prevented many bad things being built, but that it has not been notably sucessful in creating good things, or in having any overall vision of what sort of countryside it was trying to achieve. Pearce (1992) has claimed that one of its three main achievements has been the protection of the countryside, but this has produced the unintended costs of reduced economic growth, inflation of land values, and the separation of home and workplace. Finally, in a wider review, Healey and Shaw (1994) have argued that the planning agenda has been restricted to a narrow remit focused on land use and development, and to a discourse dominated by utilitarian functionalism. They conclude that if planning is to have a genuine impact on the rural environment, it needs to undergo a fundamental rethinking of its conceptions, technical methods and policy processes.

Conclusion

The main conclusion is one of ignorance in that no one really knows what impact rural planning has had on controlling the expansion of built environments into the rural areas of the United Kingdom. This is surprising in light of the enormous expense involved, not just in the payment of officials, but in the huge expenditure of time and resources by developers and pressure groups. There is thus a need for much more empirical work into the impact of planning on the ground, now there is a clearer understanding of the theory behind the processes.

IMPLEMENTATION: UNITED STATES

The pre-eminence of private property and local democracy

The United States provides a perfect contrast to the United Kingdom in terms of the three core concepts that have moulded British planning. First, the core concept that planners should be able to intervene in the use of private land has not been conceded in the United States (Krueckerberg, 1995). Indeed, the right to use land as one sees fit is enshrined in the US constitution (the Fifth Amendment states '. . . nor shall private property be taken for public use without just compensation') and attempts to take it away via planning are constantly referred to the courts under the umbrella title of the 'taking issue' (Cullingworth, 1993). Or put another way, when does the exercise of power over land use constitute such an infringement of the property right as to become a 'taking'? In the 1990s the so-called Lucas decision (Lapping and Baron, 1993) reconfirmed the primacy of private property rights. Thus, by definition, the second core concept of British planning, the non-payment of compensation for refusing or imposing conditions, cannot operate in the United States unless threats to public health, welfare or safety can be predicted by planners, allowing them to introduce zoning ordinances restricting land uses. Instead, US planners have had to invent elaborate financial systems to circumvent this major problem.

Third, the very concept of planning is not totally accepted in a country where the concepts of individual freedom, mobility and growth are deeply ingrained from both history and contemporary cultural norms. For example, Galston and Baehler (1995) argue that two tensions shape rural planning in the United States: (1) the market-oriented view of society as a whole and (2) the place-oriented view of environmentalists and planners. However, American culture gives little value to 'place' – the creation of intimate and enduring human communities. The sucess of the market-oriented culture has resulted in a mobility pattern that has undermined 'place' and the planner's aspiration to create harmonious sustainable communities. Nonetheless, Gale and Hart (1992) have found potential support for planners, albeit in the long-developed state of Maine, and with significant variations by subgroups. Beatley *et al.* (1994), from a study of 392 residents in Austin, Texas, also found some support for planning and felt that the interests of the public good must take precedence over the rights of individuals to freely use their land. Nonetheless, Kempton *et al.* (1995) have found that, although they expressed support for environmental planning, people were hypocritical since they acted differently. Furthermore, in the 1990s a significant anti-environmental planning backlash has developed which is attempting to reassert private property rights (Echevarria and Eby, 1995).

Not surprisingly, there is no federal system of planning across the United States, nor is there likely to be one in the foreseeable future. Nonetheless, most local administrations (below the state level, i.e. counties or municipalities) have developed some sort of planning system, although in reality this merely focuses on the much more limited tool of zoning and subdivision control – the division of local government areas into districts which are subject to different regulations regarding the use of land (Lapping *et al.*, 1989). Zoning is seen as an exercise of police power, under the inherent power of government, to legislate for the health, welfare and safety of the community. As such, all 50 states have passed legislation enabling local administrations to operate zoning controls for the public good, following a Supreme Court decision in 1926 which upheld the constitutionality of a zoning ordinance in the village of Euclid, Ohio (Fluck, 1986). Zoning is meant to be 'self-executing' in that permitted land uses are spelt out with such clarity there should be little room for discretion. Since there are over 83 000 local governments, control over land use operates in a very local and thus uncoordinated way, in that someone seeking to carry out a development will find a favourable zoning in at least one of these local government areas. Two broad systems for managing or preventing change to a built environment can nonetheless be discerned: farmland preservation programmes, and growth management programmes, although they are not mutually exclusive.

Farmland preservation programmes

As Table 9.1 (pages 204–205) shows, eight different programmes have tried to slow down the rate of land loss between 1949 and 1982, which tripled the amount of land devoted to urban use. Most of them centre on financial instruments, for example, differential tax assessment which taxes land as farmland, rather than the urban land it could be in its 'highest and best' use. More permanently, the purchase or transfer of development rights allows a farmer to sell the right to develop land to the government while retaining ownership of the land (Daniels, 1991). However, the incursion of new inhabitants may object to farm practices as nuisances, so right-to-farm laws have often been used to prevent incomers from imposing suburban standards. Another way to preserve farmland is to prevent incomers in the first place, by imposing minimum lot sizes via agricultural zoning policies that put such plots out of the reach of most house buyers. However, Lapping and Szedlmayer (1991) argue that most of these programmes are seriously limited by the cost implications, although more recent research by Pfeffer and Lapping (1995) has suggested that, if they were allied to comprehensive land-use planning, they could be more effective than the imperfect action they represent at present. Indeed, according to work by Nelson (1992), the experience in Oregon, where land-use planning has been well developed (Knaap, 1990), does demonstrate that prime farmland can be preserved from urbanization, a view endorsed by Coughlin (1991) from a study of land transfers in Pennsylvania.

Growth management programmes

Zoning theoretically provides a tool for growth management, but in practice it proceeds on the basis of decisions regarding individual lots (Cullingworth, 1993). In addition, the normal ethos among local administrations is that development is good, and if one person is allowed to develop then so should everybody else, on the principle of equity. Thus communities desiring to establish some sort of order in development have had to invent growth management programmes, and considerable ingenuity has been displayed in developing over 80 types of measure. Kelly (1993) has identified four broad strategies: (1) relating development to adequate infrastructure; (2) phasing growth to fit in with other objectives; (3) defining areas where development may or may not occur for set periods of time; (4) setting specific rates of growth.

A few states have also attempted to impose growth management programmes, notably Hawaii, Florida, California, Oregon, and Vermont, but each state has used a different set of techniques. Florida has limited state controls to 'areas of critical concern', whereas Oregon has set up a comprehensive system of local planning which has to conform to a long list of state goals. Gale (1992) has found they can be categorized into four paradigms: state dominant (Oregon); regional–local co-operative (Vermont); state–local negotiated (New Jersey); and a fusion model (Washington State). Bollens (1992), from a study of 13 state programmes, has found a move away from pre-emptive regulatory interventions to conjoint and cooperative state–local planning frameworks. The 1990s have witnessed renewed interest in state growth management programmes which mandate local governments to prepare and adopt comprehensive plans, and Berke and French (1994) have found this has improved plan quality.

Table 9.1 State programmes for farmland preservation

	Differential tax assessment	Centralized land-use policies	Transfer/ purchase of development rights	Agricultural zoning	Agricultural districting	Right-to-farm laws	Absentee landownership regulations and restrictions	Erosion and sediment control legislation
Alabama	✓		✓			✓		
Alaska	✓		✓			✓		
Arizona	✓				✓	✓		
Arkansas	✓					✓		
California	✓	✓	✓	✓	✓	✓	✓	
Colorado	✓		✓	✓		✓		
Connecticut	✓					✓	✓	
Delaware	✓					✓		✓
Florida	✓	✓				✓	✓	
Georgia	✓					✓		✓
Hawaii	✓	✓		✓	✓	✓	✓	✓
Idaho	✓			✓	✓	✓	✓	✓
Illinois	✓			✓		✓	✓	
Indiana	✓			✓	✓	✓	✓	
Iowa	✓			✓	✓	✓	✓	✓
Kansas	✓					✓	✓	
Kentucky	✓				✓	✓		
Louisiana	✓					✓		
Maine	✓		✓	✓	✓	✓		✓
Maryland	✓	✓	✓	✓		✓		✓
Massachusetts	✓		✓	✓		✓		
Michigan	✓			✓	✓	✓		✓

State	1	2	3	4	5	6	7
Minnesota	✓			✓	✓		✓
Mississippi	✓					✓	✓
Missouri	✓					✓	
Montana	✓				✓	✓	✓
Nebraska	✓					✓	✓
Nevada	✓				✓	✓	✓
New Hampshire	✓	✓	✓			✓	
New Jersey	✓	✓	✓	✓		✓	✓
New Mexico	✓	✓	✓	✓			✓
New York	✓			✓		✓	
North Carolina	✓	✓			✓	✓	
North Dakota	✓				✓	✓	✓
Ohio	✓					✓	
Oklahoma	✓			✓	✓	✓	
Oregon	✓	✓	✓	✓	✓	✓	✓
Pennsylvania	✓	✓		✓		✓	
Rhode Island	✓					✓	
South Carolina	✓					✓	✓
South Dakota	✓			✓		✓	✓
Tennessee	✓					✓	
Texas	✓			✓		✓	
Utah	✓			✓		✓	
Vermont	✓	✓	✓	✓	✓	✓	✓
Virginia	✓			✓		✓	
Washington	✓				✓	✓	✓
West Virginia	✓	✓	✓			✓	✓
Wisconsin	✓			✓		✓	✓
Wyoming	✓					✓	

Source: Lapping and Szedlmayer (1991), who took their data from a variety of sources. Reprinted by permission of John Wiley & Sons Ltd.

In contrast to the problems of excessive growth in some states, in other areas, notably the Midwest and the Great Plains, the problem is one of small-town decline, mainly related to the loss of farm employment, which will be exacerbated by the 1995 Farm Bill. Indeed many Americans believe that rural planning exists only in such places where the planner is a generalist and part of the community (Lapping *et al.*, 1989). In these areas a variety of economic incentives have been instituted, but Frederic (1992) has argued they have had little effect and such areas need to be linked with a metropolitan area if they are to be economically sucessful, a view empirically endorsed by Moore (1994) from a study of the Appalachians.

Conclusion

In conclusion, Cullingworth (1993) believes that planning in the United States operates, in an intolerably regressive and inefficient way, and is a servant of private interests, not the public good. A further criticism is made by Lapping *et al.* (1989), who argue that rural planning is fragmented and incremental, and is thus not able to look forward far enough to anticipate and accommodate broad economic and social changes. In terms of impact on the ground, Fuguitt (1991) has found that most of the new ex-urban settlements allowed by zoning are dominated by long-distance commuting, and Landis (1992), from a study of 10 growth control programmes in California, found they were largely irrelevant to the management of urban growth. Nonetheless, Popper (1988) is less pessimistic and argues that planning has been gaining ground constantly and will continue to gain ground. Similarly, from a study in Maryland, Feitelson (1993) found that growth controls need not necessarily be regressive or lead to greater sprawl.

In a national overview, Daniels and Lapping (1996) divide rural America into two: the core agricultural areas which are declining, and the urban fringe areas subject to massive but unsustainable expansion. In order to equalize these two areas, they argue that the United States needs more not less planning. Strong *et al.* (1996) argue that the 'taking' issue is the one that must be reformed if rural planning in the United States is to progress. Finally, in common with the United Kingdom, little is known about the detailed pattern of land-use change to the built environment, and academics have been guilty on both sides of the Atlantic of overconcentrating on theory building and process studies to the neglect of empirical fact finding, the bedrock on which theory and process studies should stand. To some extent the current state of knowledge is akin to a house of cards on a foundation of sand.

CONCLUSION

This chapter has shown how rural planning is very much a creature of the culture it finds itself in. In the United Kingdom, planning has become widely accepted because the culture is basically preservationist in a small crowded island where individual freedoms have long been surrendered for the collective good. But in the United States, the sense of space and the underlying ideology of private property have resisted attempts to impose the sort of control experienced in the United

Kingdom. Any assessment of the two systems must therefore be placed in the cultural context. From British eyes, planning in the United States looks to be a disaster; but from the US viewpoint, if their environment is a mess then they just move on out, as long as they have the cash. One day, however, even the United States will run out of places to 'mess up', then they may perceive the merits of the British planning system. Hopefully, this will not be before they have also messed up the planet with their extravagant waste of resources, as exemplified by the drive-in, throwaway, underculture of McDonalds.

In the meantime, British politicians seem to have realized just in time that the Los Angelization of England is simply not feasible, however many people have been seduced by the apparent charms of the drive-everywhere United States and demand its replication over here. Resisting their siren demands for the shallow, plastic and ersatz American environment will continue to be a major challenge for British rural planners.

In addition, it will also be a challenge for European countries which operate similar zoning systems to the United States, albeit without the impediment of the 'taking' issue (Booth, 1995). The erosion of the high density cafe society lifestyle of the European city and its replacement by low density American-style sub-urbanization represents an interesting challenge for planners seeking to manage change since, unlike the United Kingdom, continental Europe, notably France, has a massive surplus of farmland. The arguments thus centre on the need to protect the landscape heritage of both the cities and the countryside and to prevent pollution and the wasteful use of energy caused by driving everywhere. At the end of the day, the planners will have as ever to choose between the wider needs of society and the right of individuals to live as they want. This brings one full circle to the first part of this chapter, the political matrix in which planning operates; it reaffirms the need to place any study of managing change in a framework derived from Fig. 9.1 (page 190).

REFERENCES

Anderson, M. (1981) Planing policies and development control in the Sussex Downs AONB. *Town Planning Review*, **52**, 5–25.

Beatley, T., Brower, D. and Lucy, W. (1994) Representation in comprehensive planning: an analysis of the Austinplan process. *Journal of the American Planning Association*, **60**, 185–96.

Beauregard, R. (1991) Without a net: modernist planning and the post-modern abyss. *Journal of Planning Education and Research*, **10**, 189–94.

Berke, P. and French, S. (1994) The influence of state planning mandates on local plan quality. *Journal of Planning Education and Research*, **13**, 237–50.

Blacksell, M. and Gilg, A. (1982) *The countryside: planning and change*. Allen and Unwin, London.

Bollens, S. (1992) State growth management: intergovernmental frameworks and policy objectives. *Journal of the American Planning Association*, **58**, 454–66.

Booth, P. (1995) Zoning or discretionary action: certainty and responsiveness in implementing planning policy. *Journal of Planning Education and Research*, **14**, 103–12.

Brotherton, I. (1982) Development pressures and control in the national parks. *Town Planning Review*, **53**, 439–59.

Champion, T. (1993) A decade of regional and local population change: census 91. *Town and Country Planning*, **62**(3), 42–45.

Christensen, K. (1993) Teaching savvy. *Journal of Planning Education and Research*, **12**, 202–12.

Clark, G. (1992) 'Real' regulation: the administrative state. *Environment and Planning A*, **24**, 615–27.

Cloke, P. (1979) *Key settlements in rural areas*. Methuen, London.

Cloke, P. (1987) Policy and implementation decisions. In Cloke, P. (ed) *Rural planning policy into action*. Harper and Row, London, pp. 19–34.

Cloke, P. and Shaw, D. (1983) Rural settlement policies in structure plans. *Town Planning Review*, **54**, 338–54.

Cloke, P. and Thrift, N. (1987) Intra-class conflict in rural areas. *Journal of Rural Studies*, **3**, 321–34.

Cloke, P., Philo, C. and Sadler, D. (1991) *Approaching human geography: an introduction to contemporary theoretical debates*. Paul Chapman, London.

Cloke, P., Doel, M., Matless, D., Phillips, M. and Thrift, N. (1994) *Writing the rural: five cultural geographies*. Paul Chapman, London.

Coughlin, R. (1991) Formulating and evaluating agricultural zoning programs. *Journal of the American Planning Association*, **57**, 183–92.

Cullingworth, J. (1993) *The political culture of planning: American land use planning in comparative perspective*. Routledge, London.

Cullingworth, J. and Nadin, V. (1994) *Town and country planning in Britain*. Routledge, London.

Daniels, T. (1991) The purchase of development rights: preserving agricultural land and open space. *Journal of the American Planning Association*, **57**, 421–31.

Daniels, T. and Lapping, M. (1996) The two rural Americas need more, not less planning. *Journal of the American Planning Association*, **62**, 285–88.

DoE (1980) *Circular 22/80: Development control-policy and practice*. HMSO, London.

DoE (1992) *PPG7: The countryside and the rural economy*. HMSO, London.

Echevarria, J. and Eby, R. (eds) (1995) *Let the people judge: wise use and the Private Property Rights Movement*. Island Press, Washington DC.

Feitelson, E. (1993) The spatial effects of land use regulations: a missing link in growth control evaluations. *Journal of the American Planning Association*, **59**, 461–72.

Fluck, T. (1986) Euclid v. Ambler: a retrospective. *Journal of the American Planning Association*, **52**, 326–37.

Flynn, A. and Murdoch, J. (1995) Rural change, regulation and sustainability: guest editorial. *Environment and Planning A*, **27**, 1180–92.

Frederic, P. (1992) Economic development versus land use regulation. In Bowler, I., Bryant, C. and Nellis, D. (eds) *Contemporary rural systems in transition*, vol 2, *Economy and society*. CAB International, Wallingford, pp. 225–41.

Fuguitt, G. (1991) Commuting and the rural–urban hierarchy. *Journal of Rural Studies*, **7**, 459–66.

Gale, D. (1992) Eight state-sponsored growth management programs: a comparative analysis. *Journal of the American Planning Association*, **58**, 425–39.

Gale, D. and Hart, S. (1992) Public support for local comprehensive planning under statewide growth management: insights from Maine. *Journal of Planning Education and Research*, **11**, 192–205.

Galston, W. and Baehler, K. (1995) *Rural development in the United States: connecting theory, practice and possibilities.* Island Press, Washington DC.

Gilg, A. (1991) *Countryside Planning policies for the 1990s.* CAB International, Wallingford.

Gilg, A. (1996) *Countryside planning.* Routledge, London.

Gilg, A. and Kelly, M. (1996) The analysis of development control decisions: a position statement and some new insights from recent research in south-west England. *Town Planning Review*, **67**, 203–28.

Goodwin, M., Cloke, P. and Milbourne, P. (1995) Regulation theory and rural research: theorising contemporary rural change. *Environment and Planning A*, **27**, 1245–60.

Halfacree, K. (1995) Talking about rurality: social representations of the rural as expressed by residents of six English parishes. *Journal of Rural Studies*, **11**, 1–20.

Hall, P. (1974) The containment of urban England. *Geographical Journal*, **140**, 386–418.

Hall, P. (1988) The industrial revolution in reverse? *The Planner*, **74**, 15–20.

Ham, C. and Hill, M. (1994) *The policy process in the modern capitalist state.* Harvester Wheatsheaf, Brighton.

Harper, S. (1987) The rural–urban interface in England: a framework for analysis. *Transactions of the Institute of British Geographers*, **NS12**, 284–302.

Healey, P. and Shaw, T. (1994) Changing meanings of 'environment' in the British planning system. *Transactions of the Institute of British Geographers*, **NS19**, 425–38.

Held, B. and Visser, D. (1982) *Rural land uses and planning: a comparative study of the Netherlands and the United States.* Elsevier, New York.

Hill, B. and Young, N. (1991) Support policy for rural areas in England and Wales: its assessment and quantification. *Journal of Rural Studies*, **7**, 191–206.

Jackson, R. (1986) United States of America. In Patricios, N. (ed) *International handbook on land use planning.* Greenwood, Westport CT, pp. 499–526.

Jessop, B. (1990) Regulation theories in retrospect and prospect. *Economy and Society*, **19**, 153–216.

Jones, O. (1995) Lay discourses of the rural: developments and implications for rural studies. *Journal of Rural Studies*, **11**, 35–49.

Kelly, E. (1993) *Managing community growth: policies, techniques and impacts.* Praeger, Westport CT.

Kempton, W., Boster, J. and Hartley, J. (1995) *Environmental values in American culture.* MIT Press, Cambridge MA.

Knaap, G. (1990) State land use planning and inclusionary zoning: evidence from Oregon. *Journal of Planning Education and Research*, **10**, 39–46.

Krueckerberg, D. (1995) The difficult character of property: to whom do things belong? *Journal of the American Planning Association*, **61**, 301–9.

Landis, J. (1992) Do growth controls work? A new assessment. *Journal of the American Planning Association*, **58**, 489–508.

Lapping, M. and Baron, E. (1993) 1991–92, the year in review in US rural planning. *Progress in Rural Policy and Planning*, **3**, 45–56.

Lapping, M. and Szedlmayer, I. (1991) On the threshold of the 1990s: issues in US rural planning and development. *Progress in Rural Policy and Planning*, **1**, 19–35.

Lapping, M., Daniels, T. and Keller, J. (1989) *Rural planning and development in the United States.* Guilford, New York.

Lassey, W. (1977) *Planning in rural environments.* McGraw-Hill, New York.

Lindblom, C. (1977) *Politics and markets.* Basic Books, New York.

McDonald, G. (1989) Rural land use planning decisions by bargaining. *Journal of Rural Studies*, **5**, 325–35.

Marsden, T., Murdoch, J., Lowe, P., Munton, R. and Flynn, A. (1993) *Constructing the countryside*. UCL Press, London.

Moore, T. (1994) Rural planning progress in a persistent problem area: the central Appalachian example. *Progress in Rural Policy and Planning*, **4**, 17–32.

Murdoch, J. and Marsden, T. (1994) *Reconstituting rurality*. UCL Press, London.

Nelson, A. (1992) Preserving farmland in the face of urbanization: lesons from Oregon. *Journal of the American Planning Association*, **58**, 467–88.

O'Riordan, T. (1985) Future directions for enviroment policy. *Environment and Planning A*, **17**, 1431–46.

Pearce, M. (1992) The effectiveness of of the British land use planning system. *Town Planning Review*, **63**, 13–28.

Pfeffer, M. and Lapping, M. (1995) Farmland preservation, development rights and the theory of the growth machine: the views of planners. *Journal of Rural Studies*, **10**, 233–48.

Popper, F. (1988) Understanding American land use regulation since 1970. *Journal of the American Planning Association*, **54**, 291–301.

Schrader, H. (1994) Impact assessment of the EU structural funds to support regional economic development in rural areas of Germany. *Journal of Rural Studies*, **10**, 357–65.

Selman, P. (1988) Rural land-use planning – resolving the British paradox. *Journal of Rural Studies*, **4**, 277–94.

Short, J., Fleming, S. and Witt, S. (1986) *Housebuilding, planning and community action*. Routledge & Kegan Paul, London.

Sinclair, G. (1992) *The lost land: land use change in England 1945–1990*. Council for the Protection of Rural England, London.

Strong, A., Mandelker, D. and Kelly, E. (1996) Property rights and takings. *Journal of the American Association of Planners*, **62**, 5–15.

Tarrant, J. (1992) Agriculture and the state. In Bowler, I. (ed) *The geography of agriculture in developed market economies*. Longman, Harlow.

Tett, A. and Wolfe, J. (1991) Discourse analysis and city plans. *Journal of Planning Education and Research*, **10**, 195–200.

Weekly, I. (1988) Rural depopulation and counterurbanisation: a paradox. *Area*, **20**, 127–34.

CHAPTER 10

RURAL RECREATION AND TOURISM
Richard Butler

Today the temptation to escape from the turmoil of our cities into the peace and quiet of the countryside has seldom appeared more attractive and more desirable.

H. Newby, *The countryside in question*

INTRODUCTION

The use of rural areas for leisure has a long and complex history in many countries of the world, and has experienced major changes in the postwar era (Butler and Clark, 1992; Shaw and Williams, 1994). Recreation and tourism have not only reacted to global social and economic changes, but have themselves become significant agents of social and economic change in many rural areas. Indeed, one can argue that changes in rural areas relating to leisure are among the most significant to have occurred over the past three decades. Before World War II most leisure-related use of rural areas was seen, with some justification, as being of only minor importance, with relatively little direct social or economic significance to those rural areas and their populations. This is clearly not the case as the twentieth century draws to a close, for in many rural areas recreation and tourism have become the mainstays of the economy, and their existence and expansion have had profound effects upon the social and cultural patterns there.

This major shift in importance has been caused by several interrelated factors. They include a spectacular rise in participation in leisure activities in all developed countries, major changes in agriculture, significant shifts in public tastes and preferences, and the effects of a variety of technological changes and innovations. It will be argued in this chapter that, although none of these factors alone would have caused the rise in importance of recreation and tourism in rural areas, their relatively simultaneous occurrence made it almost inevitable.

Owens (1984) noted over a decade ago that much of the research on recreation and leisure in the rural context was greatly lacking in theoretical and conceptual foundations, consisting primarily of empirical studies. This situation has changed somewhat in the past decade or so, with much more effort being made to place

211

studies in a conceptual context (Bouquet and Winter, 1987; Evans and Ilbery, 1989; Butler *et al.*, 1997). Attempts to synthesize the overall nature of the effects of these activities on rural areas and their residents began with the comprehensive review of the impact literature by Mathieson and Wall (1982). Such studies as those by Gannon (1991), Getz (1986), Kousis (1989) and Reid (1995) also provide conceptually based and comprehensive assessments of the effects and implications of tourism and recreation in rural areas from differing viewpoints, reflecting the changing images and views of rural areas (Marsden *et al.*, 1993).

Before proceeding further, it is appropriate to clarify some terms. In this discussion, rural areas are taken to mean those areas lying beyond the boundaries of urban communities, but excluding extensive unsettled areas, such as the Arctic regions of Canada or much of interior Australia. Recreation and tourism are taken to be activities which are engaged in on a voluntary basis for pleasure during the participants' leisure time (Butler, 1989). Tourism is generally taken to involve greater travel than recreation, and most definitions normally involve at least one overnight stay away from home. Leisure may best be viewed as a state of mind, whereas leisure time is generally regarded as free time, i.e. time available to an individual after the normal obligations have been met.

The leisure use of rural areas can therefore be taken to include all non-essential uses of these areas, ranging from enjoying the view as one passes through to engaging in specific activities in which the rural area is a required location. In many cases, leisure use is likely to imply casual and non-specific use of rural areas primarily by local residents, and by those visiting the location for non-consumptive activities such as viewing, sightseeing, picnics, and similar relatively passive pursuits. Recreational use of rural areas has come to imply purposeful engagement in more active activities. Many recreational activities take place in a variety of settings, but can gain significantly in user satisfaction by being pursued in attractive rural locations, e.g. fishing and sailing (Shaw and Williams, 1994). Other activities may take equally attractive but different forms depending on rural or non-rural settings, e.g. visiting historic properties or dining out. However, an increasing number of recreational activities and recreational facilities are located in rural areas not because of the innate attractiveness or attributes of those areas, but because land is cheaper and more easily available than in most urban centres. It is in these urban centres that the majority of consumers reside, and access to rural areas has become increasingly easier for these urban markets.

Although the image of rural areas in many Western countries is strong and attractive (Harrison, 1991), the fact remains that many international tourists visit urban centres in greater numbers than they do rural areas. The principal attractions of countries such as Britain, France, the United States and Australia are urban, e.g. London, Paris, New York, Los Angeles and Sydney. Visits to rural locations by international tourists tend to take two forms. One is short-duration trips to specific features in the rural hinterland, especially historic sites or famed natural features (e.g. battlefields and houses, or the Grand Canyon and Milford Sound). The other is tours, often escorted on coaches, through 'marked' rural areas (e.g. the Highlands of Scotland, the Finger Lakes of New York State, or the 'standard' tour of New

Zealand described by Pearce [1995]). Personal tours of rural areas by foreign tourists are not a major feature of tourist travel, with the exception of some European countries, thus tourist exposure to most rural locations is limited and specific. The activities during tourist trips to rural areas are also more likely to be passive and the use of interpretative facilities and services is relatively high.

There are therefore a wide range of uses and values placed on rural areas in a leisure context, where leisure is taken to include recreational and tourist activities. As distinctions blur between leisure, recreation and tourism, similar to the blurring between leisure, work and retirement, it becomes more difficult to separate and differentiate the forms of pressure to which rural areas are exposed. The same locations and facilities often cater for all types of visitors, even if the perceptions and requirements of the visitors may vary widely. Traffic jams, eroded footpaths, employment and improved services occur regardless of whether visitors are engaged in leisure, recreation or tourism. The above definitions are pragmatic; they have been formulated for the purpose of the discussion which follows; more detailed discussion of definitions is contained in the literature (Bramwell, 1994; Lane, 1994a).

This chapter proceeds by discussing the ways in which recreational and tourist uses of rural areas have grown and changed over the past few decades, and provides some explanation for these changes. This is followed by a discussion of the ability of tourism and recreation to dominate many rural areas, why this trend is likely to continue, and the effects such activities are having on rural areas. The final section identifies issues and implications likely to arise from the future recreational and tourist use of rural areas, particularly in the light of changing attitudes towards rural areas generally (Cloke and Little, 1996; Murdoch and Marsden, 1994).

GROWTH AND CHANGE IN TOURISM AND RECREATION IN RURAL AREAS

Growth in participation in tourism and recreation

The growth of recreation and tourism in rural areas mirrors growth in these activities at a global scale. All of the same general forces which have affected leisure participation since the end of World War II have affected recreation and tourism in rural areas in a similar manner. These forces include increased affluence, greatly increased personal mobility, changes (often reductions) in the work schedule, greater amounts of leisure time, and technological innovations which have created greater opportunities for leisure. All of these factors and their effects on leisure have been well documented for many years (Glyptis, 1992; Jackson and Burton, 1989). They have produced two major changes in leisure participation: increased participation in most established forms of leisure; and participation in a variety of new forms of leisure, often with very rapid growth rates (Table 10.1).

Almost all activities which have traditionally taken place in rural areas have recorded significant rates of growth over the past three or four decades. And a range of new leisure activities have appeared in the past few decades, some developments

Table 10.1   Rural leisure activities

Traditional activities	New activities
Driving	Snow skiing
Walking	Snowmobiling
Visiting historic sites	Mountain biking
Picnicking	All-terrain/off-road vehicle
Nature study	Orienteering
Photography	Survival
Sightseeing	Windsurfing
Fishing	Endurance sports
Visiting rural operations	Paragliding

of existing activities, others the fruits of new technology. Many have become accessible and popular with a vast new market. Technological developments have produced lower priced and more user friendly equipment, which have allowed easier, safer and more comfortable operation than in the past. Lighter and more resilient sorts of equipment, often using 'space age' materials, have widened the availability of many activities, previously the prerogative of the rich or the superfit. The appearance of small dinghies has allowed many more people to participate in sailing than when it was a very exclusive activity based on larger boats and access to facilities.

Particularly significant is that almost all of these activities occur in rural areas. Spectacular growth in their participation has produced equally spectacular demand for access to existing and potential opportunities. Sometimes adjustments in policy have significantly increased the supply of opportunities without the creation of new facilities or major disruption of existing patterns of activity. The increasing use of reservoirs for sailing, water skiing, windsurfing and fishing is a case in point.

In other cases, informal recreational and tourist use is often made of facilities designed for other purposes. The use of forest access roads initially by walkers, and most recently by mountain bike enthusiasts, is such an example. Sometimes these new uses may present conflicts and problems to traditional and established uses, both recreational and otherwise.

In addition to new activities appearing as a result of new technology, many well-established rural leisure activities have experienced rapid growth in participation. Of particular significance are those activities related to nature and natural heritage. They reflect the rise of widespread public interest in the environment, which began in the early 1960s and was revitalized by the Green Movement from the 1980s onwards. Many of these activities occur primarily in rural areas. They include wildlife observation, collecting (fossils, plants) and nature study in general (Thrift, 1989). Bird-watching in particular is an activity which appears especially suited to rural areas because of its compatibility with other rural activities, although even this activity can create problems if participation exceeds the capacity of visited areas, communities and the birds involved.

There has also been a significant rise in popularity and demand for participation in several other leisure activities, not exclusively rural in nature or setting, but which have become increasingly located in rural areas. Golf is one of the best examples, both in demand for space and in proportional growth. Many golf courses were historically located within urban areas, or in the rural–urban fringe, but increasing numbers are now being located in rural areas, reflecting the lack of suitable land within urban areas and the fringe, and land-use planning policies. Other examples include the relocation of sports facilities, leisure centres and amusement parks to greenfield sites to take advantage of a combination of cheaper land in rural areas, highway access from a multiplicity of markets, and a more pleasant setting.

A number of activities previously enjoyed by a very limited number of people have become highly popular in recent years. This may result from (1) images created in the mass media, (2) skilful marketing and the creation of a demand, and (3) changes in tastes and fashion (Butler, 1989). Such activities commonly place an emphasis on excitement and prestige (Table 10.1), and although many are variations on well-established leisure pastimes, the way in which participation now occurs requires different, often more extensive facilities and space. Besides, current patterns of use may preclude other users from that space, even those wishing to engage in the more traditional forms of the activity. Thus free-form rock climbers shun slopes climbed by those using extensive hardware, and find new areas to climb which have not been exposed to pitons and other equipment. Invariably the desire is to have these new areas used exclusively for that specific form of recreation.

Changes in the nature of tourism and recreation

The various leisure uses of rural areas have changed in nature over time. The more traditional rural leisure activities placed an emphasis on a change of pace and setting to those experienced on a day-to-day basis in urban areas, retaining links to a simple and bucolic rural image (however mythical this may have been), and often requiring both physical and human elements in the landscape (Lane, 1994a). The new leisure uses of rural areas are much more related to the urban existence and lifestyle (Table 10.2).

There are signs here of a fundamental shift in societal preferences and behaviour. In the nineteenth century visionaries such as Frederick Law Olmstead were creating large urban parks to bring the countryside into the city (Nelson and Butler, 1974). Now the promoters of new leisure activities are bringing the urban milieu into rural areas, even though part of the participants motivation may be to escape their urban lifestyle. The countryside is increasingly becoming a setting for a wide variety of leisure activities, many of which have no specific rationale for being located in rural areas except that they require considerable expanses of land and water. Links with traditional rural leisure activities are few if any, and competition rather than complementarity with traditional rural activities is common. That there is not more negative reaction from permanent rural residents to the appearance of these activities is initially surprising, but reflects a number of items discussed in more detail below, including the opportunity to earn money from such activities,

Table 10.2   Characteristics of rural leisure activities

Traditional activities	New activities
Relaxing	Individualistic
Family/group oriented	Competitive
Non-competitive	Active
Passive	High cost
Non-mechanized	Relatively high per capita impact
Rural landscape complementary	Mechanized
Rural land uses complementary	High technology
Low cost	Prestigious
Low per capita impact	Fast paced
Low technology	Rural landscape irrelevant
Non-urban	Rural land uses competitive
Minimum skill or training required	Urban related
	Skill demanding

the change in ownership and function of much of the rural hinterland in many areas, and apparent general governmental indifference to, or even encouragement of, these changes (Curry, 1992; Langford, 1994; Ministry of Tourism and Recreation, 1986; Tourism New South Wales, 1994). Besides, there are fewer traditional rural residents left in many rural areas to voice such a protest.

CHANGES IN ORIGIN AND DESTINATION AREAS

Some of the reasons for the expansion of leisure use of rural areas can be found in the changes which have taken place in society at large during the same period. It is important to note that these changes reflect both changes in the demand and in the supply sides of the leisure equation. Changes in rural areas can both help and hinder the growth in leisure activities there, as do changes in the origin areas and their populations.

First and foremost are increases in global population numbers. If nothing else had changed over the postwar period, the increases in overall population would inevitably have translated into increased use of rural areas for leisure activities. Increases in disposable incomes have occurred for many people in developed countries over most of this period. A reasonable proportion of this income has been allocated to expenditure on leisure-related items, including equipment, accommodation, transport, access/membership fees to facilities and training. Many of these acquisitions are interdependent; the purchasing of equipment may lead to the purchase of training and access to facilities, and also to accommodation on site and improved transport for accessing the site. Once such investments have been made, continued and possibly accelerated growth in participation can be expected, if only to justify the initial expenditures. Venturing into rural areas to engage in such activities on a frequent and regular basis may expose the user to other opportunities that may also be taken up.

Another factor is the changes in taste and fashion that have taken place in Western societies. Especially among more educated and affluent segments of society, increasing attention has been paid to fitness and health (Cordell *et al.*, 1985). This has been mirrored by increased participation in related activities, including walking, swimming and cross-country skiing, all primarily rural activities. Allied to these activities has been growth in more glamorous activities such as downhill skiing, water skiing, sailing, canoeing, spelunking, horse riding, and competitive activities, which though they require varying amounts of physical effort, also assist in the cultivation of a desired image.

Working in conjunction with these desires is the increased interest in nature and things green. Rural areas are perceived, rightly or wrongly, as more natural than urban areas, and are thus felt to be inherently more appropriate for recreation, indeed for all aspects of living. Spectacular growth in various forms of ecotourism is part of the result of this interest, although motives on both the demand and supply sides may not be as pure as they initially appear (Wheeller, 1993). In addition, activities which could occur in urban areas assume greater significance and appear to give greater satisfaction in rural settings, e.g. shopping, particularly for rural boutique items, wine and foodstuffs.

Added to this set of changes should be included increasing dissatisfaction with urban environments, which are seen as dirty, unhealthy, crowded, stress filled, dangerous and generally unattractive. Rural areas are increasingly viewed as safe, healthy, offering a cleaner, less stressful setting for living, raising a family and enjoying leisure. Those who can afford it now attempt to have the best of both worlds by purchasing a rural retreat in which they can recreate and recharge themselves at weekends and in holiday periods (Cherry, 1993). Those who cannot afford a property of their own can utilize timeshare or condominium arrangements in purpose-built properties, increasingly numerous in attractive rural areas, or visit somewhere such as a Centre Parc, the up-market equivalent of holiday camps (Croall, 1995). Still others who have the occupations which allow it, are moving permanently to rural areas and either commuting to urban employment or conducting their business from their new rural locations. Modern telecommunications and computer facilities are making this arrangement easier and more common. These new rural residents are often more active in terms of participation in leisure activities and more affluent than traditional rural residents. Part of their rationale for moving to rural areas may have been the desire to be able to participate in recreational activities more frequently, as well as moving to what is felt to be a more attractive living environment.

Technological improvements and innovations have spurred the growth in the use of rural areas for recreation and tourism in several ways (Countryside Commission, 1992). The effects of such development upon the activities themselves have already been noted. Another major impact is in improved access to and within rural areas. Although access by public transport has generally declined in most rural areas in developed countries as rail and bus services have been reduced, access by private transport has increased in many areas. In all developed countries, car ownership has increased during the postwar years which, with improved highway provision, has provided easier access to rural areas for much of the urban population (Page, 1995). The use of technology for one activity may also lead to increased ease of access and participation in other, sometimes unrelated activities. One example is the provision of lift facilities for downhill skiing; in the summer season they can be used to provide access to hills for many more people than would otherwise visit such locations. The provision of chairlift access to the top of Cairngorm in Scotland from the early 1960s (Watson, 1984) has allowed large numbers of summer visitors to reach a vulnerable environment previously inaccessible except to sturdy walkers. This situation is mirrored in many other ski areas in Europe, North America, Australia and New Zealand.

The great increases in leisure use of rural areas has also been facilitated by the process of rural restructuring, discussed elsewhere in this book. The decline of the traditional agricultural population, tied to what were once family farms, is a common process in most Western countries (Bowler *et al.*, 1992). Rural depopulation and other factors such as changes in agricultural production methods, mechanization, agribusiness practices and set-aside policies have reduced the political power and influence of rural areas, weakened the economic and social viability of communities and associated services, and changed the physical appearance of the countryside. These changes have taken place in all rural settings, from the traditional long-settled countryside of Europe to what were the frontiers of North America and Australia less than a century ago (Nelson *et al.*, 1975).

At the same time, a growing segment of the urban population of these countries has been keen and able to utilize rural areas for leisure. One result has been the purchase of rural properties for leisure and related purposes, normally taking them out of agricultural use on a permanent basis (Clarke, 1993; Coppock, 1977; Denman, 1978; Halseth and Rosenberg, 1995). Hobby farms, second homes, commuter homes and retirement homes are individual forms of this process; and country estate developments, timeshare projects, leisure complexes, holiday estates and retirement communities are more commercial versions of the same phenomenon (Butler, 1985; Gartner, 1987; Stroud, 1983). The urban searcher for rural real estate, driven by amenity and ludic desires, can frequently outbid potential local purchasers. This can result in social and political concern, economic and social change and alterations in spatial and temporal patterns of land use (Wyckoff, 1990).

Along with this process of urban acquisition of rural areas, there has been a fundamental change in the perception and image of rural areas, from locations for agriculture, forestry and water production, to sites for leisure activities and developments (Cloke, 1993). Rural areas have shifted from being areas of primary production to settings for quaternary service industries, particularly those in recreation and tourism (Curle and Rounds, 1995). This change is reflected in many rural municipalities facing problems of new ex-urban residents complaining about basic agricultural practices, a significant change from earlier times when rural residents complained of urban intrusions into the countryside.

Somewhat paradoxically, there has also been an increasing demand for things rural, and by implication, green, sustainable, traditional, and environmentally appropriate (Bramwell, 1994; Denman, 1994). Thus the production and sale of foodstuffs, artefacts and crafts in rural areas finds a ready market among the urban residents and the temporary rural leisure residents. Small wineries in many rural areas have become attractions for visitors to the countryside, both as sites for visits and for purchases. So appealing has this activity become that winery tours are now available on a commercial basis from cities such as Canberra and Niagara Falls, and increasing numbers of wineries have established visitor services, including restaurants and gift shops (Hall and Macionis, 1997). In Scotland, whisky distilleries are a logical and very successful counterpart to wineries, and many distilleries now employ more staff as guides and in restaurants and in gift shops than they employ in the production of the whisky. Distillery visits are appearing with increasing frequency on tour itineraries (McBoyle, 1994), and the marketing of whisky and wine trails has been undertaken by several tourist boards in Britain, Canada and Australia.

A related form of rural leisure development is the appearance of festivals. Some of them are even related to traditional rural activities, successors of traditional harvest festivals, and many are spectacularly successful (Janiskee and Drews, 1997). The Elmira Maple Syrup Festival in Ontario has grown from a small festival attracting hundreds of people each year in the 1960s to an event drawing over 20 000 in 1995. Maple syrup has declined in importance as attendance at the festival has increased, alongside an increase in some of the other attractions. In Ontario there are now some 363 annual festivals, many of which have rural themes and timetable their schedules so that regions have festivals on consecutive weekends through certain periods of the year (Butler and Smale, 1991).

THE CHANGING IMAGE OF RURAL AREAS

Another factor which has significantly affected the use of rural areas for leisure purposes has been the changing image of these areas in the public's mind (Murdoch and Marsden, 1994). Rural areas, particularly those most closely mirroring the bucolic image of the countryside (Harrison, 1991), have always been attractive for urban residents to visit for relatively passive and non-consumptive leisure uses. In recent years this image has been reinforced and modified through the deliberate utilization of rural areas by the entertainment media. One may suggest that many urban residents' experience of rural areas is at best fleeting and increasingly brief, although perhaps it is becoming more frequent. That is, urban citizens may visit rural areas more often than in the past, but the visits are not to explore or experience the rural milieu in any depth, and visitors' familiarity with and knowledge of these areas is extremely limited. The inclusion of attractive and evocative images of rural areas into popular media presentations appears to stimulate and create a demand to visit the landscapes they portray (Butler, 1990, ch. 3).

The images of the American West have been the backdrop for many Western movies, although these images do not appear to have attracted great numbers of sightseeing tourists in themselves, perhaps because the emphasis in these films has traditionally been on action rather than scenery. One exception is the use of Monument Valley and its spectacular landforms in movies from the original *Stagecoach* onwards. More recently, the successful movie *Dances with Wolves*, much of which was filmed in South Dakota, has spawned several small companies catering for tourists wishing to visit the landscape so effectively portrayed in the film (Riley, 1994). Much of the evidence is anecdotal on how landscapes appearing in film and television affect visits to the locations shown (Tooke and Baker, 1996). However, in the North American context, movies such as *Witness* (portraying the Amish country of Pennsylvania), *Close Encounters of the Third Kind* (filmed at Devil's Tower National Monument in Wyoming), *Field of Dreams* (set in an Iowa cornfield) and, most recently, *The Bridges of Madison County* all appear to have stimulated considerable interest by sightseers desiring to see the locations used in the films (Riley, 1994).

In the Canadian situation, one rural area which has benefited economically from literary and visual imagery is Prince Edward Island. The island has been popularized by the writings of L.M. Montgomery in the *Anne of Green Gables* stories (Squire, 1992), particularly popular with Japanese teenage girls. The area surrounding the principal setting of the stories is dominated by tourist development, some related to the image and much simply capitalizing on the name alone. Irrespective of the appropriateness of the development, the prime attraction of Prince Edward Island is the rural image promoted in the books and subsequent television series and video productions. Herbert (1996) also discusses the appeal of artistic and literary places in France which have become tourist attractions.

In a different environment, the success of Australian films such as *The Man from Snowy River*, its sequel and the associated television series have portrayed attractive images of the Australian countryside which have drawn visitors to the region. The Australian film *Crocodile Dundee* and its sequel, however, are widely credited with creating a whole new image for the Australian outback and its residents; combined with other events, the films appear to have revitalized the Australian overseas tourism industry and drawn a considerable number of visitors to Australia (Riley, 1992).

It is the English countryside, however, which remains pre-eminent in providing a backdrop to films and television series that attract large numbers of sightseers. When combined with the literary image of the British countryside, the effect is extremely powerful (Butler, 1990). Films such as *Howards End*, *The Remains of the Day*, *The Shooting Party*, *Room with a View*, and *Sense and Sensibility* have provided a series of visual images of rural and social nostalgia of great effectiveness. These images have been heightened by an equally popular series of television programmes, ranging from *Brideshead Revisited* to *Heartbeat*, and from *The Forsyte Saga* and *Pride and Prejudice* to *Wycliffe*. The mix of historical aristocratic life in impressive surroundings and contemporary dramas in attractive rural settings has stimulated a significant desire to visit the locations of these programmes (Tooke and Baker, 1996).

So much interest has been generated that tourist boards have published maps showing the location of sites used during filming. The power of the camera to attract visitors to a site is revealed in a book discussing four such films, Pym's *Merchant Ivory's English Landscape: rooms, views and Anglo-Saxon attitudes* (1995). In the flyleaf it says that the book 'furnishes travelers to England with gazetteers telling how to visit the places where the films were shot, and including such key information as travel directions from London, opening hours of houses and gardens, descriptions of other nearby places of interest'. It remains to be seen whether recent successful films featuring other parts of rural Britain (*Braveheart* and *Rob Roy*) will broaden their appeal in the same way.

The influence of films and television on shaping an image of rural areas and subsequently increasing the numbers of visitors has not always been accepted by local residents as beneficial. Film makers sometimes have to relocate because of the numbers of tourists who descend on their locations, as happened in the case of the producers of the television series *Heartbeat*, set in a Yorkshire village. Some local residents find the portrayal of their home area and subsequent appearance of visitors undesirable and steps have been taken to limit the continued use of specific locations, as has happened in the village of Pennant, in Scotland, used in the film *Local Hero*.

The portrayal and subsequent marketing of such rural areas and the true effects on leisure visits have been little researched, but this process is likely to become more significant in the future for several reasons. The visual medium appears to be more influential to current and probably future generations, which appear to read less than earlier generations and thus rely more on what they see to influence them. Many levels of government, aware of the financial rewards from film and television production, aggressively pursue production companies to have films made in their locations for the economic returns (Riley, 1994). Rural locations which have maintained their traditional rural image are likely to remain in high demand in the future and become even more significant as tourist attractions, especially as more and more rural locations become developed and lose their mythical image.

CHANGING ATTITUDES TO TOURISM AND RECREATION IN RURAL AREAS

Compounding the problems discussed above are the changing attitudes of many rural residents to additional leisure use of rural areas (Allen *et al.*, 1987; Cordell *et al.*, 1993; Rothman, 1978). Access to private land for public leisure activites has long been a de facto right in many countries and a legal right in others, such as Scandinavia. Increases in the numbers of potential users, the incidence of inappropriate behaviour and the mechanization of activities, however, have made many rural residents less willing to allow casual access to their properties (Cordell *et al.*, 1985). In North America, in particular, issues of landowner liability and changes in legislation in recent years have made all landowners increasingly concerned about allowing visitors access to their properties. A variety of alternative approaches have been tried, including intermediary association, disclaimers of liability and signposting of prohibited activities (Butler and Troughton, 1985; Kaiser and Wright, 1985; Wright and Fesenmaier, 1988).

It has been shown that the attitude of some new, especially ex-urban, residents in rural areas is more negative to the use of their property by others than the attitude of long-time residents. The result is that some areas which were freely accessible on an informal and unofficial basis are no longer accessible to the general public (Lee and Kreutzwiser, 1982). Changes in attitudes towards hunting in North America have seen its proponents face significant problems in finding areas to hunt. Familiarity with, and acceptance of, hunting as a traditional rural activity is lower among recent ex-urban rural residents than among long-term rural residents, and few are willing to allow access to their properties for hunting by anyone. In the future, fishermen may find the same reluctance among new rural landowners to allow them access to rivers and lakes. The new residents may wish to preserve the fishing for themselves, or they may see no reason to condone the killing of fish for sport.

A similar problem has developed in the case of water bodies in Canada and the United States, where second homes now completely ring some popular lakes. The result is that the public is effectively denied access to these public water bodies and the lakes become de facto private facilities because the shoreline has been alienated from the public domain.

At present, the less consumptive activities such as walking, nature observation and photography have escaped some of the negative reaction that more active and consumptive activities have faced, presumably because they cause less disturbance than other activities and have a long history in many rural areas. In the future, however, the sheer numbers of potential participants is likely to mean that even such relatively passive activities face increasing restrictions on the privately owned areas in which they are allowed (Wyckoff, 1990). Although the tradition of public access to private rural areas has a long and firmly established tradition in Europe, even there the traditional rights of way are being challenged and closed, or made increasingly difficult to use (Countryside Commission, 1986, 1994a, 1995b).

If present trends continue, it is extremely likely that participants in tourism and recreation in rural areas will be limited to engaging in many of their activities on a decreasing amount of land, primarily land which is within the public domain (Cordell *et al.*, 1993). Although vast expanses of such land still exist, especially in North America, much of it may not be free of limitations on use. Logging, grazing and mineral rights may have been leased to private enterprise, and such land is often very considerable distances from most potential users. Access to rural areas and communities may remain easy for the urban motorist, but access to rural land and water for active leisure activities may become increasingly limited to formal public and private facilities. Such facilities will almost inevitably involve admittance charges, considerable numbers of users and a loss of spontaneity and privacy. The opportunities for a casual trip to a rural area for a quiet walk in the country are likely to be increasingly rare if present trends continue.

THE EFFECTS OF TOURISM AND RECREATION ON RURAL AREAS

The impacts of tourism and recreation on destination areas have been the focus of a great deal of research for he past two decades (Mathieson and Wall, 1982; Pearce, 1989; Smith and Eadington, 1993). Much of the empirical research in earlier years dealt with the economic impacts of tourism and recreation, to provide a justification for the involvement of communities in catering for tourists and recreationists, and much of this was conducted in rural areas and settlements (Broadbent, 1972; Davies, 1971; Hoyland, 1982; Vogeler, 1977). This was followed by a shift in emphasis to the environmental effects of tourist and recreational activities; however, much of this research was carried out in wild or unsettled areas, such as national parks or wilderness areas, where the protection of natural environments was of particular importance. Relatively little research was conducted on environmental impacts in what may be regarded as conventional rural areas. Some of the work undertaken by, and on behalf of, the Countryside Commission and the former Nature Conservancy Council in the United Kingdom does deal with environmental concerns (Countryside Commission, 1991, 1993, 1995d). Increasing attention has been paid in the past two decades to the social and cultural effects of tourism on populations of destination areas and their reactions to tourism and recreation (Allen *et al.*, 1987; Bouquet and Winters, 1987; Brougham and Butler, 1977; Rothman, 1978).

It is clear that the impacts of leisure activities vary considerably with the intended role and the actual role played by tourism and recreation. To contemplate tourism and recreation as only being capable of minor effects and always destined to play a supplementary role in rural communities and economies is to severely underestimate the potential effects of these activities. The leisure industry is larger at the global scale than agriculture and more important in terms of international trade than almost anything else (Pearce, 1989). The negative and undesired effects of these activities in destination areas often result because the magnitude of the leisure industry in all its forms is not well understood or investigated before development occurs. Once developed, tourism and recreation have a pattern of taking over communities and economies, and moving from a supplementary or complementary role to one of domination (Butler, 1991). Investigation of areas such as Catalonia in Spain (Hermans, 1981), Strathspey in Scotland (Getz, 1986), and central Florida surrounding the Disney complex reveals this quite clearly (Walsh, 1992).

In many rural areas the environmental effects of tourism and recreation tend to be localized, often confined to specific public recreation sites, such as national, provincial and country parks. In these areas the effects can be severe, even irreparable, as dramatically noted by Croall (1995). To most rural residents, however, the more serious impacts are economic and social (Allen *et al.*, 1987; Brougham and Butler, 1981). Most visitors to rural areas do not do extensive damage to the agricultural environment, but the sheer weight of numbers of users at specific popular sites can cause major environmental damage. Such pressure of numbers can also have deleterious effects on human features of the landscape, as has been the case at Stonehenge in England and the Lascaux Caves in France. A disturbing feature is that, with the advent of more sophisticated equipment and the effects of other social and economic changes noted earlier, leisure users of rural areas are venturing in increasing numbers into areas which had previously been spared extensive use. The Cairngorm Plateau in Scotland, backcountry trails in the Rockies, and the Australian outback have all received rapid increases in visitor numbers in the past two decades, often with clearly identified impacts and change.

As greater interest is being expressed in nature-related activities, in heritage and ecotourism, the pressure on remoter, and often more vulnerable, rural areas can only be expected to continue. If the undesirable effects of such use are to be contained and minimized, then proactive planning and development will be required far more frequently and effectively than has mostly been the case up to now.

POLICIES TOWARDS TOURISM AND RECREATION IN RURAL AREAS

Until recently the growth of tourism and recreation in most rural areas had taken place in a policy vacuum. When tourism and recreation were considered in policy development, they were almost always viewed as minor issues not worthy of specific attention and tended to be dealt with dismissively, or else viewed as one of a number of possible economic Band-aid solutions to rural marginality and decline. Before the late 1960s, rural tourism and recreation were never treated as major land uses with potential economic and social effects. Only when these activities began to be perceived as causing difficulties, such as land-use conflicts, overcrowding and environmental degradation, did public sector agencies begin to take steps to alleviate them through policy and planning. Even now, however, some policy statements and proposals for rural areas still neglect or underplay the role and effects of tourism and recreation (Scottish Office, 1995).

Few countries have been as active in countryside planning as the United Kingdom, with the establishment of the Countryside Commission in the late 1960s and attempts to provide policy and management guidance to both the public and private sectors. From their inception, however, such bodies have faced the problem of not having ownership of the areas over which they had responsibility. Despite this, the Countryside Commission has produced innovative and often successful management schemes and approaches to deal with the problem of some 600 million visits a year to the British countryside (Sharpley, 1993). But their advice, and that of many other voluntary and statutory bodies in the United Kingdom, has often been disregarded or has at best been implemented in a piecemeal and poorly supported manner. As a result, although positive management actions have created successful attractions and features, the absence of necessary funding for operations and maintenance has increased the pressure from newly generated demand. Sometimes this has aggravated the pressures on both existing and new facilities. Innovations in policy and management, such as long-distance footpaths, integration of public and private transport to provide access to areas, and upland management schemes, have demonstrated what effective research and development can achieve, but lack of permanent funding for operation remains a problem.

In other countries, approaches have been somewhat different, largely occasioned by the fact that ownership of many of the prime attractions and much of the land has been in public sector control since the time of settlement. Thus in Canada, the United States, New Zealand and Australia, national, provincial and state parks provide vast areas of high quality scenery and many recreational and tourist opportunities, which are managed by the public sector under common policies and guidelines for each system. There are disadvantages to such arrangements. One is the development of an island mentality; that is, the focus of efforts in management and policy on the islands (parks, recreation areas, campgrounds and other facilities) to the exclusion of the surrounding countryside in which they are located. Although the special places are protected to varying degrees, they also become the focus of attraction, and most park systems in the countries listed above are now experiencing overuse and environmental impacts. These problems are not new; they were observed and predicted in the US national park system almost three decades ago (Darling and Eichorn, 1967). Only in recent years have agencies begun to incorporate surrounding areas into their planning policies and actions, and to involve other agencies in developing strategies to handle the increasing numbers of visitors to these areas.

It is important to recognize that the appreciation and desire to preserve the image of the countryside exist to varying degrees in different countries. The long-settled European countries tend to have strong positive feelings about preserving a living rural heritage, both in terms of the aesthetic and the cultural elements of the countryside. They view it as a scenic and leisure resource of considerable importance. In more recently settled countries, such attitudes are less strong, and more emphasis has traditionally been given to preserving unsettled wild lands, with change and development accepted as the norm for the settled countryside (Troughton *et al.*, 1975). Only recently has pressure been exerted to preserve and maintain rural areas on non-economic grounds (Byrnes, 1995), as well as to maintain a rural and agricultural population.

Policies on tourism and recreation in rural areas, where they have existed, have often exacerbated the division between tourism and recreation, instead of treating them as part of the same phenomenon – the use of rural areas for leisure. Thus, policies focused on recreation tend to favour provision and management of opportunities, many of them in the public sector, in part as meeting a social need and in some cases reflecting concerns over impacts, particularly environmental ones (Countryside Commission, 1991, 1994b). But because of perceived economic gains, policies concerning tourism in rural areas are often focused upon the stimulation of visitor numbers. Restrictions on visitor numbers and activities are rarely discussed; the emphasis is normally on marketing the potential of rural areas for tourism (Gilbert, 1989) and ensuring means of access for visitors and opportunities for them when they arrive.

Despite the belated acknowledgement of the economic potential of tourism, some governments at the national level still ignore it, assuming or ensuring it will be developed by the private sector, and their efforts tend to be confined to marketing and very limited research (Pearce, 1992). In the past two years the US government has abandoned a federal presence in tourism, and involvement by the Canadian and UK governments has decreased significantly over the past decade. Although other countries place more importance upon tourism, few have specific policies relating to tourism in rural areas. Australia has the most specific policy, a formal document on rural tourism (Commonwealth Department of Tourism, 1995) based on a series of earlier studies (Australian Tourism Industry Association, 1991; Australian Tourist Commission, 1983; Commonwealth Department of Tourism, 1992, 1993, 1994; Morison, 1995). In other countries, and at other levels, rural tourism has been developed, but often in the absence of any formal policy, and sometimes by a variety of agencies with different goals and objectives. In Ontario, as noted by Reid *et al.*, 'The subject of rural tourism . . . is relatively new in its current form' (1993, preface). Although tourism has existed for a long time, formal recognition, policies and planning have appeared only recently.

Much of the research specifically on rural tourism and recreation dates back no more than two decades (Broadbent, 1972; Chow, 1980; Countryside Commission, 1981; Dernoi, 1983; Edmunds, 1974; Middleton, 1982; Vogeler, 1977). Most of this research has concentrated on case studies and possible avenues of development rather than policy issues; Hoyland (1982) was one exception to this pattern. Only in the past few years has some research dealt with the development of possible policies for tourism in rural areas (Countryside Commission, 1993; Luloff *et al.*, 1994), and with examples of development and policy implications (Australian Tourism Industry Association, 1991; Clarke, 1993, Chon and Evans, 1989). One aspect which has received particular emphasis is the linking of tourism in rural areas with sustainable development of tourism (Bramwell, 1994; Croall, 1995; Countryside Commission, 1995a,c; Denman, 1994; Lane 1994b). The logic of linking rural tourism with sustainable development may contain a large element of wishful thinking. Agriculture is increasingly unsustainable in many areas, and tourism has never distinguished itself as being either sustainable or taking a long-term view of development. It does, perhaps, reflect increasing awareness that tourism and, by implication, recreation need to be appropriate in the scale and nature of their development, particularly in areas with existing populations and patterns of economic activity.

Other agencies recognized this fact (Government of Alberta, 1985; Wight, 1988) before sustainable development became fashionable, although policies and actions have not always matched this recognition. It is hard to argue with the sentiment expressed by Hill, who defined a value system for rural tourism as including 'a commitment to sustainability, careful tourism planning, regional cooperation, quality service, hospitality, and preservation of artificial, cultural and natural environments' (1993, p. 123). The problem is that few policies for rural tourism exist and fewer still incorporate rural recreation. None seem able to guarantee the results will be either sustainable or compatible with existing economic, social and ecological processes already established in those rural areas. Without successfully combining tourism, recreation and leisure, then planning for the integration of these elements with existing and proposed other uses for rural areas, any policy is likely to be at best short-lived and will probably achieve only limited success.

FUTURE ISSUES AND IMPLICATIONS

The responsibility for protecting rural areas has always been difficult, irrespective of the countries involved (Gilg, 1978; Curry, 1992; Byrnes, 1995). In earlier years that task was somewhat easier than it is now because of the increasingly complex mix of uses and perceptions of rural areas. The overwhelming perception was that rural areas were primarily for food production plus other minor uses, and there was little argument that agricultural production was the appropriate dominant use which should be protected and supported where necessary. Now rural areas are seen in a variety of different ways. The multiplicity of interest groups insist on establishing and preserving what they claim to be their rights in rural areas. Thus not only is there an agricultural lobby, but also in most rural areas there are support groups for forestry, for environmental issues, for non-farming residents, for recreation and tourism users, for industries and for heritage protection. In certain areas, various other groups may be added to this list, including the military, other public sector agencies and departments, water authorities and neighbouring urban authorities. Perhaps of almost equal significance is the fact that few of these groups are themselves unified in their desires and perceptions about the rural areas they wish to access.

In the context of tourism and recreation, competition is often as fierce between interests as it is between tourism/recreation and other interests. Those requiring access to mountains and hills for walking are often adamantly opposed to ski developments (as is the case in several areas in Scotland). Sailboat enthusiasts are frequently opposed to the presence of water-skiers and powerboat enthusiasts on the same body of water (as in the Lake District in England). Anglers dislike canoeists or kayakers, walkers of trails are opposed to horse riders and trailbike riders on the same trails (in Ontario and in Alberta), and cross-country skiers dislike snow-mobilers in the same environment (almost anywhere in North America). Those wishing to use the countryside for quiet relaxation are annoyed by the development of theme parks, many forms of aerial recreation and rock festivals.

As long as each lobby group or special interest group is continuing to demand its rights to rural areas, there is little hope of achieving contentment in those areas, as Frost and Jay (1967) commented three decades ago. As Curry (1992) notes, achieving integration, even within the limited context of recreation, access, amenity and scientific conservation, has been relatively unsuccessful in the United Kingdom despite considerable efforts and several reports. With the continued blurring between recreation, tourism and amenity living (including retirement), the leisure-related demands on rural areas are likely to continue to increase, both in number and complexity. It matters little if the perceptions of these user groups about of rural areas are neither accurate nor appropriate (Harrison, 1991).

Tourism, recreation and leisure represent uses of rural areas which are growing rapidly. They exercise considerable influence in these areas because of the numbers of users involved and the financial expenditures generated. Present and future land-owners and decision makers cannot avoid taking them into account as major factors to be considered in planning and developing rural areas in the twenty-first century (Shaw and Williams, 1994). Although the past two millennia have seen rural areas as being dominated by food production, the next millennium may see them being dominated by leisure activity.

REFERENCES

Allen, L.R., Long, P.T. and Perdue, R.R. (1987) The impact of tourism development on residents' perceptions of community life. *Journal of Travel Research*, **XXVII**(3), 16–21.

Australian Tourism Industry Association (1991) *Development of farm and country holidays in Australia*. Australian Tourism Industry Association, Canberra.

Australian Tourist Commission (1983) *Farm holidays*. Australian Tourist Commission, Melbourne.

Bouquet, M. and Winter, M. (eds) (1987) *Who from their labours rest: conflict and practice in rural tourism*. Gower, Aldershot.

Bowler, I.R., Bryant, C.R. and Nellis, M.D. (eds) (1992) *Contemporary rural systems in transition*, vol. 2 CAB International, Wallingford.

Bramwell, B. (1994) Rural tourism and sustainable rural tourism. *Journal of Sustainable Tourism*, **2**(1/2), 1–6.

Broadbent, K.P. (ed) (1972) *Tourism and recreation in rural areas*. Commonwealth Bureau of Agricultural Economics, Oxford.

Brougham, J.E. and Butler, R.W. (1977) *The social and cultural impact of tourism: a case study of Sleat, Isle of Skye*. Scottish Tourist Board, Edinburgh.

Brougham, J.E. and Butler, R.W. (1981) A segmentation analysis of resident attitudes to the social impact of tourism. *Annals of Tourism Research*, **8**(4), 569–90.

Butler, R.W. (1985) Timesharing: the implications of an alternative to the conventional cottage. *Leisure and Society*, **VIII**(2), 769–80.

Butler, R.W. (1989) Tourism and tourism research. In Jackson, E. and Butron, T.L. (eds) *Understanding leisure and recreation*. Venture Publishing, State College PA, pp. 567–96.

Butler, R.W. (1990) The influence of the media in shaping international tourism patterns. *Tourism Recreation Research*, **15**(2), 46–55.

Butler, R.W. (1991) Tourism, environment and sustainable development. *Environmental Conservation*, **18**(3), 201–9.

Butler, R.W. and Clark, G. (1992) Tourism in rural areas: Canada and the United Kingdom. In Bowler, I.R., Bryant, C.R. and Nellis, M.D. (eds) *Contemporary rural systems in transition*, vol 2. CAB International, Wallingford, pp. 166–86.

Butler, R.W. and Smale, B.J.S. (1991) Geographical perspectives on festivals in Ontario. *Journal of Applied Recreation Research*, **16**(1), 3–23.

Butler, R.W., Hall, C.M. and Jenkins, J.M. (eds) (1997) *Tourism and recreation in rural areas*, John Wiley and Sons, Chichester.

Butler, R.W. and Troughton, M.J. (1985) *Public use of private land*. University of Western Ontario, London.

Byrnes, B. (1995) *Saving the countryside – conserving rural character in the countryside of southern Ontario*. Conservation Council of Ontario, Toronto.

Cherry, G.E. (1993) Changing social attitudes towards leisure and the countryside. 1890–1990. in Glyptis, S. (ed) *Leisure and the environment*. Belhaven, London, pp. 22–32.

Chon, K.-S. and Evans, M.R. (1989) Tourism in a rural area – a coal mining-county experience. *Tourism Management*, **10**(4), 312–21.

Chow, W.T. (1980) Integrating tourism with rural development. *Annals of Tourism Research*, **7**(4), 584–607.

Clarke, J. (1993) *An analysis of the demand for farm tourism accommodation in Scotland*. Scottish Tourist Board, Edinburgh.

Cloke, P. and Little, J. (eds) (1996) *Contested countryside cultures*. Routledge, London.

Cloke, P.J. (1993) The countryside as commodity: new spaces for rural leisure. In Glyptis, S. (ed) *Leisure and the environment*. Belhaven, London, 53–70.

Commonwealth Department of Tourism (1992) *Tourism – Australia's passport to growth*. Department of Tourism, Canberra.

Commonwealth Department of Tourism (1993) *Rural tourism*, Discussion Paper 1. Australian Government Printing Service, Canberra.

Commonwealth Department of Tourism (1994) *National rural tourism strategy*. Australian Government Printing Service, Canberra.

Commonwealth Department of Tourism (1995) *Rural tourism in Australia*. Government Printing Services, Canberra.

Coppock, J.T. (ed) (1977) *Second homes – blessing or curse?* Pergamon, London.

Cordell, H.K., Gramman, T.H., Albrecht, D.E., McLellan, R.W. and Winthrow, S. (1985) Trends in recreational access to private rural lands. In *Proceedings of the National Outdoor Recreation Trends Symposium*, Myrtle Beach SC. US Department of Interior, National Parks Service, vol 1, pp. 164–84.

Cordell, H.K., English, D.B.K. and Randall, S.A. (1993) *Effects of subdivision and access restrictions on private land recreation opportunities*, General Technical Report RM231. Rocky Mountain Forest and Range Experimental Station, USDA. Forest Service, Fort Collins.

Countryside Commission (1981) *The public on the farm*. John Dower House, Cheltenham.

Countryside Commission (1986) *Access to the countryside for recreation and sport*, CCP 217. Countryside Commission, Cheltenham.

Countryside Commission (1991) *Visitors to the countryside*, Consultation Paper 341. Countryside Commission, Cheltenham.

Countryside Commission (1992) *Trends in transport and the countryside*, CCP 382. Countryside Commission, Cheltenham.

Countryside Commission (1993) *Principles for tourism in the countryside*, CCP 429. Countryside Commission, Cheltenham.

Countryside Commission (1994a) *Managing public access*, CCP 450. Countryside Commission, Cheltenham.

Countryside Commission (1994b) *Working for the countryside 1994/1995*. Countryside Commission, Cheltenham.

Countryside Commission (1995a) *Action guide to sustainable rural tourism*. Countryside Commission, Cheltenham.

Countryside Commission (1995b) *National rights of way conditions survey 1993/94*. Countryside Commission, Cheltenham.

Countryside Commission (1995c) *Principles of sustainable rural tourism*. Countryside Commission, Cheltenham.

Countryside Commission (1995d) *The environmental impact of leisure activities in the English countryside*. Countryside Commission, Cheltenham.

Croall, J. (1995) *Preserve or destroy tourism and the environment*. Calouste Gulbenkian Foundation, London.

Curle, R. and Rounds, R.C. (1995) *A Manitoba case study: impact of non-resident land ownership and permanent recreational development in traditional farm areas*. Rural Development Institute, Brandon.

Curry, N. (1992) Recreation, access, amenity and conservation in the United Kingdom: the failure of integration. In Bowler, I. R., Bryant, C.R. and Nellis, M.D. (eds) *Contemporary rural systems in transition*, vol 2. CAB International, Wallingford, pp. 141–54.

Darling, F.F. and Eichorn, N.D. (1967) *Man and nature in the national park*. Conservation Foundation, Washington DC.

Davies, E.T. (1971) *Farm tourism in Cornwall and Devon, some economic and physical considerations*. Report 184, Agricultural Economics Unit, University of Exeter.

Denman, R. (1978) *Recreation and tourism on farms, crofts and estates*. A report to the Highlands and Islands Development Board and the Scottish Tourist Board. Scottish Tourist Board, Edinburgh.

Denman, R. (1994) Green tourism and farming. In Fladmark, J.M. (ed) *Cultural tourism*. Donhead, Wimbledon, pp. 215–22.

Dernoi, L. (1983) Farm tourism in Europe. *Tourism Management*, **4**(3), 155–66.

Edmunds, H. (1974) Farming and tourism. *Journal of the Royal Agricultural Society of England*, **135**, 42–49.

Evans, N.J. and Ilbery, B.W. (1989) A conceptual framework for investigating farm-based accommodation and tourism in Britain. *Journal of Rural Studies*, **5**, 257–66.

Frost, D. and Jay, A. (1967) *To England with love*, Hodder, London.

Gannon, A. (1991) Rural tourism as a factor in rural community economic development for economies in transition. *Journal of Sustainable Tourism*, **2**(1/2), 51–60.

Gartner, W.C. (1987) Environmental impacts of recreational home developments. *Annals of Tourism Research*, **14**(1), 38–57.

Getz, D. (1986) Tourism and population change: long-term impacts of tourism in the Badinoch and Strathspey district of the Scottish Highlands. *Scottish Geographical Magazine*, **102**(3), 113–25.

Gilbert, D. (1989) Rural tourism and marketing. *Tourism Management*, **16**(1), 39–50.

Gilg, A.W. (1978) *Countryside planning – the first three decades*. Methuen, London.

Glyptis, S. (1992) The changing demand for countryside recreation. In Bowler, I.R., Bryant, C.R. and Nellis, M.D. (eds) *Contemporary Rural Systems in Transition*, vol 2. CAB International, Wallingford, pp. 155–65.

Government of Alberta (1985) *Position and policy statement on tourism*. Policy Statement 1. Government of Alberta, Edmonton.

Hall, C.M. and Macionis, N. (1997) Wine tourism in Australia and New Zealand. In Butler, R.W., Hall, C.M. and Jenkins, J.M. (eds) *Tourism and recreation in rural areas*. John Wiley, London.

Halseth, G. and Rosenberg, M.W. (1990) Conversion of recreational residences: a case study of its measurement and management. *Canadian Journal of Regional Science*, **13**, 99–115.

Halseth, G. and Rosenberg, M.W. (1995) Cottages in an urban field. *Professional Geographer*, **47**(2), 148–59.

Harrison, G. (1991) *Countryside recreation in a changing society*. TMS Partnership, London.

Herbert, D.T. (1996) Artistic and literary places in France as tourist attractions. *Tourism Management*, **17**(2), 77–86.

Hermans, D. (1981) The encounter of agriculture and tourism: a Catalan case. *Annals of Tourism Research*, **8**(3), 462–79.

Hill, B.J. (1993) The future of rural tourism. *Parks and Recreation*, **Sept**, 98–101, 123.

Hoyland, I. (1982) The development of farm tourism in the UK and Europe: some management and economic aspects. *Farm Management (UK)*, **1**(10), 383–89.

Jackson, E.C. and Burton, T.L. (1989) *Understanding leisure and recreation.* Venture Publishing, State College PA.

Janiskee, R.L. and Drews, P.L. (1997) Rural fesitvals and community reimaging. In Butler, R.W., Hall, C.M. and Jenkins, J.M. (eds) *Tourism and recreation in rural areas.* John Wiley, London.

Kaiser, R.A. and Wright, B.A. (1985) Recreational access to private land: beyond the liability hurdle. *Journal of Soil and Water Conservation*, **40**(6), 478–81.

Kousis, M. (1989) Tourism and the family in a rural Cretan community. *Annals of Tourism Research*, **16**(3), 318–32.

Lane, B. (1994a) What is rural tourism? *Journal of Sustainable Tourism*, **2**(1/2), 7–21.

Lane, B. (1994b) Sustainable rural tourism strategies: a tool for development and conservation. *Journal of Sustainable Tourism*, **2**(1/2), 102–11.

Langford, S.V. (1994) Attitudes and perceptions toward tourism and rural regional development. *Journal of Travel Research*, **XXXII**(3), 35–43.

Lee, A.G. and Kreutzwiser, R. (1982) Rural landowner attitudes towards sportfishing access along the Saugeen and Credit rivers, southern Ontario. *Recreational Research Review*, **9**, 7–14.

Luloff, A.E., Bridges, J.C., Graefe, A.R., Saylor, M., Martin, K. and Gitelson, R. (1994) Assessing rural tourism efforts in the United States. *Annals of Tourism Research*, **21**(1), 46–64.

McBoyle, G. (1994) Industry's contribution to Scottish tourism: the example of malt whisky distilleries. In Seaton, A.V. (ed) *Tourism – state of the art.* John Wiley, Chichester, pp. 517–28.

Marsden, T., Murdoch J., Lowe, P., Murton, R. and Flynn, A. (1993) *Constructing the countryside.* UCL Press, London.

Mathieson, A. and Wall, G. (1982) *Tourism: economic, social and physical impacts.* Longman, Harlow.

Middleton, V.T.C. (1982) Tourism in rural areas. *Tourism Management*, **3**(1), 52–58.

Ministry of Municipal Affairs (1986) *Planned retirement communities.* Province of Ontario, Toronto.

Ministry of Tourism and Recreation (1986) *The recreation road – a rural route to planning.* Province of Ontario, Toronto.

Morison, J.B. (1995) *Rural tourism needs analysis.* Australian Rural Management Services, Parkside SA.

Murdoch, J. and Marsden, T. (1994) *Reconstituting rurality: class, community and power in the development process.* UCL Press, London.

Nelson, J.G. and Butler, R.W. (1974) Recreation and the environment. In Manners, I.R. and Mikesell, M.W. (eds) *Perspectives on Environment.* Association of American Geographers, Washington, pp. 290–310.

Nelson, J.G., Troughton, M.J. and Brown, S. (1975) *The countryside in Ontario.* University of Western Ontario, London.

Owens, P.L. (1984) Rural leisure and recreation research: a retrospective evaluation. *Progress in Human Geography*, **8**(2), 157–88.

Pearce, D.G. (1989) *Tourist development.* Longman, Harlow.

Pearce, D.G. (1992) *Tourist organisations*. Longman, Harlow.

Pearce, D.G. (1995) *Tourism today: a geographical analysis*. Longman, Harlow.

Pym, J. (1995) *Merchant Ivory's English Landscape: rooms, views and Anglo-Saxon attitudes*. Abrams, New York.

Reid, D.C. (1995) Tourism: saviour or false hope for the rural economy? *Plan Canada*, **May**, 22–26.

Reid, D.G., Fuller, A.M., Haywood, K.M. and Bryden, J. (1993) *The integration of tourism, culture and recreation in rural Ontario: a rural visitation program*. Ministry of Agriculture, Toronto.

Riley, R. (1992) Movies as a tourism promotion: a 'pull' factor in a 'push' location. *Tourism Management*, **13**(3), 267–74.

Riley, R. (1994) Movie induced tourism. In Seaton, A.V. (ed) *Tourism – state of the art*. John Wiley, Chichester, pp. 453–58.

Rothman, R.A. (1978) Residents and transients: community reaction to seasonal visitors. *Journal of Travel Research*, **XVI**(4), 8–13.

Scottish Office (1995) *Rural framework*. Scottish Office, Edinburgh.

Sharpley, R. (1993) *Tourism and leisure in the countryside*. ELM, Huntingdon.

Shaw, G. and Williams, A.M. (1994) *Critical issues in tourism – a geographical perspective*. Blackwell, Oxford.

Smith, V.L. and Eadington, W.R. (1993) *Tourism alternatives, potentials and problems in the development of tourism*. University of Pennsylvania Press, Philadelphia PA.

Squire, S.J. (1992) Ways of seeing, ways of being: literature, place and tourism in L.M. Montgomery's Prince Edward Island. In Simpson-Housley, P. and Norcliffe, G. (eds) *A few acres of snow: literary and artistic images of Canada*. Dundurn, Toronto, pp. 134–47.

Stroud, H.B. (1983) Environmental problems associated with large recreational subdivisions. *Professional Geographer*, **35**(3), 303–13.

Thrift, N. (1989) Images of social change. In Hammett, C., McDowell, L. and Sarre, P. (eds) *The changing social structure*. Sage, London.

Tooke, N. and Baker, M. (1996) Seeing is believing: the effect of film on visitor numbers to screened locations. *Tourism Management*, **17**(2), 87–94.

Tourism New South Wales (1994) *Regional tourism strategy*. Tourism New South Wales, Sydney.

Troughton, M.J., Nelson, J.G. and Brown, S. (eds) (1975) *The countryside in Ontario*. University of Western Ontario, London.

Vogeler, I. (1977) Farm and ranch vacationing. *Journal of Leisure Research*, **9**(4), 291–300.

Walsh, D.J. (1992) The evolution of the Disney Land environs. *Tourism Recreation Research*, **XVII**(1), 33–47.

Watson, A. (1984) Paths and people in the Cairngorms. *Scottish Geographical Magazine*, **100**(3), 151–60.

Wight, P. (1988) *Tourism in Alberta*. Environment Council of Alberta, Edmonton.

Wright, B.A. and Fesenmaier, D.R. (1988) Modeling rural landowners' hunters' access policies in East Texas. *Environmental Management*, **12**(2), 229–36.

Wyckoff, W.K. (1990) Landscapes of private power and wealth. In Conzen, M.P. (ed) *The making of the American landscape*. Harper Collins, London, pp. 335–54.

CHAPTER 11

SERVICE PROVISION AND SOCIAL DEPRIVATION

Owen Furuseth

INTRODUCTION

For most residents of the industrialized world, who live in urban and suburban communities, the term *rural* conveys a comfortable image of picturesque small towns and open countryside populated by prosperous farmers and other middle-class or similar residents. Rural areas represent an idyllic community: an open and clean environment, free of the stress and the pathologies associated with fast-paced urban living, simple face-to-face relationships and neighbourliness, and a local economy that thrives on nature's abundance and hard work. In rural areas, one imagines the best that life has to offer.

Missing from these bucolic images are decaying farm homesteads, the small clusters of houses or mobile homes in disrepair, emptying towns with vacant store-fronts and crumbling infrastructure, and the abandoned factories, mines and mills. These elements are also an important part of the rural landscape of the developed nations and they are the physical evidence of a marginalized population, the rural poor and disadvantaged. The poor have always been a largely invisible part of the rural environment, but recently, who they are, their numbers and their relationship with their neighbours and outsiders have been changing.

The debates over welfare reform and discussions over an underclass culture of inner city neighbourhoods make it is clear that the public and elected officials seem fixated on poverty as an urban issue. Ignored in the dialogue are issues and policy questions related to property and deprivation beyond the city's edge. Throughout the industrialized nations, geographically focused rural poverty is extensive, with significant numbers of rural residents facing chronic poverty and deprivation. In the United States there were 7.6 million rural people living in poverty, or 13% of the total rural population in 1990. The incidence of poverty in rural America is higher for almost all racial or ethnic groups. It is even higher than in central cities. Rural Americans have higher rates of unemployment and mortality and less access to education and employment training and other human services that urban Americans take for granted (Wimberly, 1993). In Britain the scope of rural poverty and deprivation is less pervasive but still significant. Using relative income status data, Bradley (1986) estimated 3.2 million rural residents could be classified as poor. As a result, he concluded that relative poverty should be considered a severe social problem in village England. Similar findings are reported from Australia, where rates of poverty and unemployment are higher in rural areas and the range of work opportunities is much narrower (Lawrence, 1990).

The evidence suggests that the rural poor and the problems associated with poverty and deprivation in the countryside were a more salient public issue in the 1950s and 1960s (President's National Advisory Commission on Rural Poverty, 1967). In the United States and the United Kingdom several blue ribbon government panels reported on the scope of rural employment, infrastructure and social problems, and made recommendations for addressing them. But most of the serious policy recommendations were never carried out. Indeed, during the past two decades, the plight of the rural poor has largely escaped the attention of opinion leaders and policy makers. In increasing urban societies, pockets of chronically poor and disadvantaged rural people have become more invisible.

The purpose of this chapter is to extend the framework laid out above by examining the effects of rural transformation on public services and rural population. Particular emphasis is placed on the nature and context of rural poverty and deprivation, and the relationship between social condition and service provision. The scope and complexity of issues related to services and service provision in rural areas are too broad to cover completely. Accordingly, the chapter focuses on the conceptual issues revolving around two selected service areas: housing and health care. The purpose is to illustrate the general relationships that exist between service provision and social status through these specific examples.

POLITICAL AND ECONOMIC RESTRUCTURING

Among the most powerful trends shaping social and economic life in the latter part of this century has been the restructuring of the global economy and rise of neoconservative political regimes in Western industrial nations. The impacts of the restructuring processes have altered different levels of social process and relationships between groups and locales. Discourse surrounding the effects of the restructuring of global and national economies has generally been cast in terms of disruptions in urban labour markets and urban-focused regions. In a similar fashion, conservative public service reforms are seen as disproportionately impacting disadvantaged inner city communities. What have generally gone unrecorded are the impacts of these economic and political changes on the countryside, in particular the rural poor. In agriculturally dependent areas, mechanization and land consolidation have weakened social relations between classes and reduced the economic viability of rural service centres. In other industrial and resource-based rural areas, job losses related to restructuring processes and tightened environmental controls have exacerbated the long-standing problems of intermittent employment and outmigration. For some observers, these social impacts are inevitable externalities accompanying changing economic conditions. But to others, the beneficial economic impacts are far outweighed by the erosive social and economic effects on local rural communities.

In the conservative political climate that has emerged in the Western nations, the reduction in financial assistance and the implementation of libertarian policy directions have impacted the provision of services and infrastructure to rural areas disproportionately. The swing of the political pendulum to the right has seen the adoption of more conservative fiscal programmes and the enactment of a wide array of market-based public policies in areas from agricultural supports to transport and social service funding. In this new political climate, local governing bodies are given more responsibility and fewer resources. As a consequence, disadvantaged rural communities compete against urban and other rural jurisdictions for a shrinking resource pool in a contest in which they have no expectation of winning. Thus, although total government expenditures to social services, housing and public services have dropped in absolute terms, the proportional impact of the decrease has been greatest on poor rural communities.

The scramble for government assistance is part of a larger place-based competition aimed at securing mobile public and private capital (Peck and Tickell, 1994). Rural governments increasingly find themselves in a domestic and international competition to attract new capital investments. For example, a small town in rural Atlantic Canada may easily find itself trying to hold on to a local industrial plant that is weighing relocation offers from another small town in the south-east United States or an industrial park in Central America. The risk of this new local competition is that uneven development between places and regions will be increased, with poor rural areas further marginalized by wealthier, stronger growth areas (Sengenberger, 1993). Empirical evidence of increasing disparity in economic opportunity between disadvantaged rural regions in the United States and suburban and urban counties has already been well documented by Falk and Lyson (1988).

UNEVEN SOCIAL DEVELOPMENT

A central construct of this chapter will be uneven social development. The notion of unevenness in social relations is an extension of the theory of uneven economic development and derives from inequality in economic and social opportunity. The restructuring processes of the recent past have accelerated the spatial pattern of uneven social development between rural and urban areas. But these principal outcomes are far from universal. Some rural areas have enjoyed economic benefits from these changes. Indeed, they have been repopulated and have adapted well to the economic and political shifts. However, other rural areas have been irreparably affected. The result is an increasing social differentiation in rural areas and a widening unevenness in rural social geography. In response, established reciprocal relationships are giving way, relationships between rural locales that formerly served to construct a 'rural interest' and guard it in the political arena. The interests of prosperous rural areas have less in common with disadvantaged rural communities and more in common with suburban interests. Thus, on one level, this chapter reports on social unevenness between rural and urban areas and the effects that it has on shaping rural life. Equally important, however, are the impacts of internal unevenness, the growing gap between rural areas, which adds weight to the debate surrounding what is rural (Halfacree, 1993). For if the term *rural* is to have meaning, it must on some level distinguish between urban and rural areas. Increasingly, *rural* in the Western, industrial nations is not crafted by traditional economic and social variables, but by synergetic notions rooted in geographical relationships and community.

ISSUES IN RURAL SERVICE PROVISION

Rural communities require essential public services in a quantity and quality that is equivalent to urban areas. Aside from the egalitarian argument for geographical equity in access to public services, rural servicing is critical to community sustainability. A full array of community services is universally recognized as a fundamental ingredient in rural development programmes. Nonetheless, rural–urban discrepancies in service provision and standards have evolved over time and persisted throughout the world (Lonsdale and Enyedi, 1984). A 1991 report of the Organization for Economic Cooperation and Development (OECD) concluded that rural service delivery in OECD countries was deteriorating.

In general, rural services can be classified by a four-part typology derived by scale and user groups (Organization for Economic Cooperation and Development, 1991). First come services intended to open up rural areas; that is, service provision designed to make rural areas more accessible to the larger world. The most powerful tools in this category are communication networks. Second are the basic infrastructure services which encompass the primary array of services necessary to support human development, such as water supplies, electricity and roadways. Third come services designed to enhance the quality of life, a second tier of infrastructure that represents an additional level of more costly public service. Although they may be perceived as basic to rural residents, government increasingly views them as betterments, beyond the scope of basic service. Examples include expanded educational and health care facilities, postal services and recreation. Fourth are services to business, including consultancy services, research and development (R&D) investments and upgraded infrastructure that provide a platform for rural business interests. Service delivery and the need for government 'subsidy' of service provision vary between rural areas according to local conditions.

Even under ideal circumstances there are economic barriers to providing adequate community services in rural locales. The demographic reality of rurality means dispersed populations and low relative population density. The consequence is that the potential demand for services delivered from discrete facilities is dispersed and the per capita costs of providing services are higher than in urban or suburban settings with greater population densities. Depending upon who is paying for service provision, there is a tendency either to scale back or not to provide the service based upon an assessment of the cost and need for the services. For example, fire protection and library services may be unavailable, whereas health care and public education may be partially provided or supplied in a less costly fashion.

Rural service reforms

At a time when conservative political regimes are increasingly dominant in the Western industrial nations, and the political clout of rural areas is declining, the restructuring of rural services has emerged as a widespread integral element of central government reform. Over the past 20 years, a reduction in rural service provision has been a significant component of conservative government reform in Britain, Australia, Canada, New Zealand and the United States. The scale and scope of the restructuring have varied, but it is generally marked by three strategies: (1) the privatization and corporatization of state organizations or departments, (2) the withdrawal of state support for community-based services or infrastructure, and (3) the devolution of government responsibility for service provision to local government units. Among the specific rural service delivery strategies implemented in OECD countries have been the creation or consolidation of subregional service centres (Austria, France, Sweden), locally based self-help programmes (United Kingdom, Finland, Sweden), segmentation of service location based upon life-cycle stages, joint public/private services (United Kingdom, France, Finland) and mobile or peripatetic services (France, Australia, United States) (Organization for Economic Cooperation and Development, 1991).

The underlying assumption of these reforms is the necessity to reduce the size and cost of the public sector, focusing foremost on the most economically inefficient activities. Fitchen (1991) has labelled this the 'cost-effectiveness model' for rural service provision. Because of the application of economic rationality, rural communities are more likely to experience the effects of these reforms than urban and suburban jurisdictions. As a part of this thrust, the private market and local governments are considered to be better managers and therefore more capable of making difficult policy choices. They have therefore been given responsibility over what were formerly central government assets and jurisdictional authority.

Impact on economically distressed areas

The impacts of service provision restructuring on rural jurisdictions vary with geographical setting and social context. In isolated rural areas or rural communities that are either economically distressed or stagnating, the withdrawal or reduction of government services and support is harsh. Wimberly (1993) warns that inadequate infrastructure in economically troubled rural areas represents one of the most serious threats to social and economic development. Existing rural institutions and businesses are often unable to respond to the 'opportunity' or 'challenge' of diminished support from government. Outmigration is fostered, especially among younger workers and more affluent residents, and the proportions of service-dependent populations are increased, leading to further declines in the local quality of life. In many cases, rural residents tend to blame themselves for the problems rather than looking at the larger political and economic processes. When the median age of the community reaches approximately 35, deaths begin to outnumber births and decline becomes as much a function of low birth rates as outmigration (Jackle and Wilson,

1992). A declining population base and greater impoverishment do little to reverse declining infrastructural conditions.

The experiences in parts of rural New Zealand illustrate the destabilizing effects of restructuring on distressed rural areas. Among the areas most severely impacted by the conservative reforms begun in 1984 was the West Coast, the remote natural resource-based region on the South Island. Sparsely populated and isolated from economic and population centres, the West Coast was reliant on state support for maintaining the socio-economic fabric. Pawson and Scott (1992) have recorded the economic and social disruption that have buffeted the region in the aftermath of government reforms. The impacts have included higher unemployment in both the private and public sectors; outmigration, especially by younger people; declining levels of public services; and increased levels of social unrest and community conflict, often focused on agencies or groups thought to be threatening the wellbeing of the region. The cumulative effect of the reforms has been to challenge 'the right to live, learn, and work' (Hudson and Sadler, 1986, p. 182). The choices left to West Coasters are either to leave, protest or acquiesce (Hirschman, 1970). These experiences are not unique. They are shared by many other rural areas that are ill-equipped or unable to make the transition accompanying government service reforms.

In the face of sustained rural outmigration and costly service provision, some commentators in the United States have advocated a 'rural triage' for those increasingly marginalized small towns and communities (Daniels and Lapping, 1987). Under this model, infrastructure and service provision would selectively be invested in areas that could be expected to remain economically viable. These communities would act as growth poles. Conversely, those locales not chosen would receive no new infrastructure or expanded services, and would be expected to be abandoned.

The application of the triage concept has been most widely discussed with reference to the Great Plains region, including portions of eight states extending from North Dakota to Oklahoma and west to Montana and Colorado. This large farming and ranching region has consistently been losing population and employment. Because of the seemingly irreversible decline, Popper and Popper (1987) have advocated human retreat from the region and the creation of a 'Buffalo Commons', an enormous greenline park, where natural processes would be allowed to operate and human development would be restricted to environmentally friendly activities. Other rural areas that have similar development projections and could be likely candidates for rural triage include northern New England and parts of West Virginia and eastern Kentucky, in central Appalachia.

At this point, the notion of rural triage is controversial and presents little political support. However, one might wonder whether the current political and institutional framework that reduces service provision to distressed rural areas is not already implementing a non-spatial, incremental triage.

Impacts on prosperous areas

For rural areas experiencing economic growth and in-migration from outside the area, the issues relating to government restructuring and service provision are quite different. In this context there are likely to be conflicts and social cleavage over how and where the effects of government reforms will be implemented. Population and economic growth translate into demand for new and increased public service and infrastructure. The costs of expanding services must be borne by local residents. Moreover, new residents coming from urban or suburban settings expect a wider range and/or a more in-depth set of services and infrastructure than long-time rural residents. Also, newcomers often hold different land-use and development policy priorities, leading to political tensions. Former urban residents are likely to be more 'green' in their politics than traditional residents. The utilitarian beliefs of long-time residents run counter to the 'preservationist' goals of newcomers and are played out in local land-use and environmental policy making.

Changing government involvement in service provision has stretched the fiscal capacity of rural communities to accommodate new growth. Community cohesion or agreement with respect to public service provision and priorities is diminished and the cleavage and conflict between new and old residents is heightened. Within the community, the people most likely to be adversely affected by these circumstances are poor and disadvantaged residents. They operate from a position of powerlessness and their status within the community is further diminished by the change. In California, Bradshaw (1993) has observed that rural communities experiencing rapid growth and economic diversification have undergone bifurcation. Newcomers and those long-time residents able to accommodate economic and social change exert control over the local government apparatus. In turn, they steer public policy and services toward their vision of community development. Revisions in local government services and infrastructure programmes tend to increase local property taxes as well as imposing new land development and environmental regulations. All of these changes diminish the capacity of the poor to better their status. The poor are left poorer and socially marginalized. Fieldwork carried out in Britain (Marsden *et al.*, 1993) and Canada (Bunce and Walker, 1992) has reported similar conclusions.

Although the direct impacts of decreased state support for service provision vary across the rural landscape, depending upon the economic status and social relations in the community, the process is fundamentally disruptive to the rural community. It weakens social institutions and creates greater localized social conflict. Within rural areas the groups most likely to be underserved as a result of cutbacks are the most immobile populations, those unable to travel to service provision points, and populations residing in the most geographically remote locations. These populations generally comprise the rural poor and disadvantaged.

RURAL DEPRIVATION

Social and economic inequalities that are evidenced between urban and rural communities as well as among rural jurisdictions have attracted increasing discussion of rural deprivation. Rural deprivation is generally presented in terms of a lack of resources that are considered fundamental to a normal life and work. Indeed,

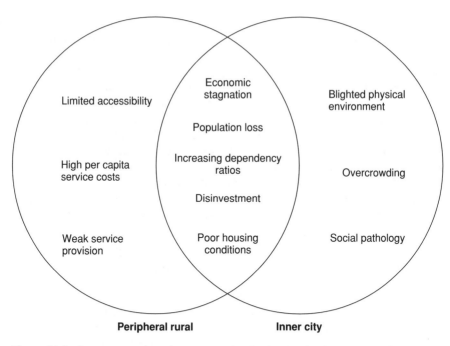

Figure 11.1   Interconnectedness between rural and urban deprivation (Adapted from Moseley, 1980; Pacione, 1984)

deprivation derives from a resource-based framework emphasizing a lack of accessibility and employment opportunities, inadequate housing, marginal service provision, and lack of social and public services. Shaw (1979) identified three related types of rural deprivation: household (income and housing), opportunity (employment, education, health and recreation) and mobility (accessibility and transport costs).

There is also a qualitative measure of deprivation linked to social conditions, measured by impacts on family ties, social cohesion, and community stability and organizations. Lewis (1983) has called this immaterial deprivation and suggested that it has greater significance than resource-defined deprivation. Poverty is relative and can be perceived as a temporary condition; deprivation is to feel not a part of society (Eyles, 1987). A lack of connectedness to societal integration is manifested by higher rates of social problems, weak community structures, as well as chronic economic problems. Combining the two measures presents a more fully developed explanation for deprivation. Thus, an individual's perceptions of deprivation are constructed from resource inadequacies as well as people's experiences and expectations of social disconnectedness.

Rural deprivation is often linked to specific rural at-risk populations: the elderly, the disabled, children and minorities. However, larger-scale patterns of spatial inequality are also evident. The localization of deprivation in rural areas is evident in widely disparate locations, from the Pacific to Europe and North America. What might seem odd is the commonality that exists between deprived rural and inner city communities (Fig. 11.1). Both of these areas are economically and politically peripheral to policy-makers and a majority of citizens in their respective nations. For rural communities, problems of geographical isolation and/or inaccessibility foster economic and institutional barriers to breaking out of a deprivation trap; whereas inner urban areas are not spatially isolated, instead they are socially and politically isolated by racial and cultural differences, and fear of crime and other social problems in inner city areas.

Deprived rural and urban areas share common problems (Moseley, 1980). The recent economic restructuring has afflicted both areas strongly, leading to higher levels of unemployment and reducing the local government revenue base. In turn, service delivery needs have been increased by the economic decline, disproportionately affecting service-dependent populations, triggering a spiralling deterioration in government services and quality of life, and spurring further outmigration and economic disinvestment. Indeed, the economic context of deprivation facing Black residents of inner city London or Los Angeles is shared with rural residents of either Appalachia or Ireland.

Theories of rural deprivation

Two very different conceptual theories have been advanced for explaining deprivation in rural Britain. Although they are built on small-scale empirical research in Britain, and therefore reflect a particular institutional and social context, both have application to other industrial nations (Lowe *et al.*, 1986).

Sociological explanation A sociologically oriented explanation draws heavily on the seminal field studies of Newby and colleagues. It explains rural deprivation in the context of the process of rural repopulation marked by the replacement of the traditional farming community with a new community of non-agricultural ex-urbanities, often retirees and commuting workers. The newcomers are generally affluent and cause a reallocation of political and economic power and social status. Customary social relations are disrupted and low income, low status members of the community are further marginalized. This social shift leads to changes in service provision, local economic structure and general social amenities. Resource-related deprivation among the rural working class and the poor is exacerbated by the higher living expenses driven by increased costs of living and higher tax rates. Concurrently, the change in community structure leads to deepening qualitative deprivation, altering traditional social relations and subordinating disadvantaged populations. The sense of knowing one's place in the community and belonging to the community is diminished.

Planning explanation An alternative theory of rural deprivation is posited in work by Moseley (1979) and Shaw (1979). Labelled a planning theory of deprivation, it draws heavily on locational theory. This paradigm is far more focused on a resource-centred concept of deprivation. Using a central place framework, rural deprivation is seen to be caused by distance from employment and service opportunities. Relative geographical isolation imposes economic and social costs on immobile populations. Differential accessibility leads to intracommunity deprivation or broader place-based deprivation.

In this context, deprivation is conditioned by decision-making processes beyond the local level. Government planning schemes and locational decisions by private firms which affect employment and service provision can enhance or reduce the opportunity for betterment. It follows, therefore, that planning authorities using their control over spatial development patterns and infrastructure possess the opportunity to distribute resources equally and to steer private capital investment. This would create enhanced accessibility, increased rural opportunity and diminished deprivation.

Moseley and Shaw point out that state planning organizations can operate in reverse, causing a 'planned deprivation' (McLaughlin, 1986). Increasingly, rural services are pulled back as decision-making processes are focused upon collective resource demands and the higher costs of providing rural services. The most geographically isolated rural communities and politically impotent rural populations receive the least consideration and become more inaccessible. Rural land-use planning processes and environmental conservation are seen as contributing to the problem, not simply because they ignore the distributional equity questions, but because they support the hegemonic control of the property-owning class against the needs of the poor and disadvantaged.

THE RURAL POOR

The urban poor are a highly visible fixture of life in the industrialized world. They live in well-defined inner city neighbourhoods, and their faces and voices appear on the local and national news media often in association with stories about violence and crime. They are the subject of growing debates over an underclass subculture. The rural poor, on the other hand, are becoming more invisible. The existence of an underclass culture is controversial, with critics on the intellectual left and right. It remains a social theory that warrants discussion and attention. Ironically, the underclass thesis developed out of a rural context, but currently it is not widely examined by rural social scientists.

Beginning in the 1950s, sociologists and anthropologists began to apply the conceptual framework of rural poverty to urban areas. Although there are important differences between the work of Banfield (1958), Harrington (1962), Myrdal (1963), Lewis (1966) and others, a common theme was the unique values, attitudes and lifestyle of a growing poverty class. This underclass culture, as it has come to be called in the professional literature and popular media, consists of populations that have very weak or non-existent attachment to the labour market and rely on welfare and underground economic activities (e.g. crime) to survive (Kelso, 1994). Furthermore, gender is a marker for the underclass, with single mothers representing a large proportion (Garfinkel and McClanahan, 1994). Among families headed by single mothers, 71% in the United Kingdom, 39% in Canada and Germany, and 38% in the United States are dependent on government benefits for more than half their income. The urban underclass is disproportionately composed of minorities and is geographically concentrated in declining urban neighbourhoods.

Widespread application of the underclass definition to rural social geography has been limited. There is no universal evidence to suggest the underclass is a significant element of rural poverty in any industrial nation. Cloke (1992) has noted how the scope of rural poverty in Britain is likely to be unreported and has raised questions about an underclass culture in the countryside. In research examining multidimensional deprivation in Scotland, Pacione (1995) has explored similar themes, noting the complex weaving of economic, social and geographic variables which lead to pervasive, chronic deprivation in parts of urban and rural Scotland. Some North American social scientists have applied the underclass culture to selected rural Aboriginal communities in Canada and the United States (Sandefur, 1989). Duncan (1992) and Fitchen (1981) have argued that a rural underclass has developed in isolated districts of the southern Appalachian Mountain region and northern New York. But neither researcher has extended the concept to the larger rural geography.

What is common to both the urban and rural underclass literature is the absence of 'social buffers' and community resources which can soften the effects of poverty and foster the escape from an environment that traps poor people (Mincy, 1994). For the rural underclass, geographic isolation and lack of access to economic and educational resources are fundamental barriers to improvement. Consequently, the best chance for escaping the rural underclass is migration. As Duncan (1992, p. 131) observes, 'the poor only move up by moving out.'

Rural poverty versus urban poverty

In contrast to the urban poor, rural poverty is generally invisible to most observers. This imperceptibility occurs for several reasons. Lowe *et al.* (1986) argue that rural poverty in Britain has remained a non-issue, owing to the power of agricultural interests to perpetuate the rural idyll as a description of rural life. In this cultural context, the city is the repository of the poor, and the countryside is home of the affluent and self-sufficient.

An alternative explanation suggests the rural poor are less conspicuous because they are a traditional segment of the larger rural community. The poor are camouflaged by social and cultural conformity. They are neither spatially concentrated nor segregated. Rather, they live and work in the midst of more affluent neighbours. The rural poor are also less likely to be 'different' from others. In the United States, Canada, Australia and New Zealand, higher proportions of black, brown and native peoples live in poverty; however, poverty does not follow racial or ethnic boundaries and large numbers of majority populations are also disadvantaged.

Invisibility aside, the large numbers of rural poor make statistical documentation of their presence unavoidable. For example, the geography of poverty in the United States is largely shaped by rurality (Fig. 11.2). Large concentrations of poor counties, i.e. counties with at least 20% of the residents living below the poverty line, are found in a broad swath across the South and the South-West. Central Appalachia,

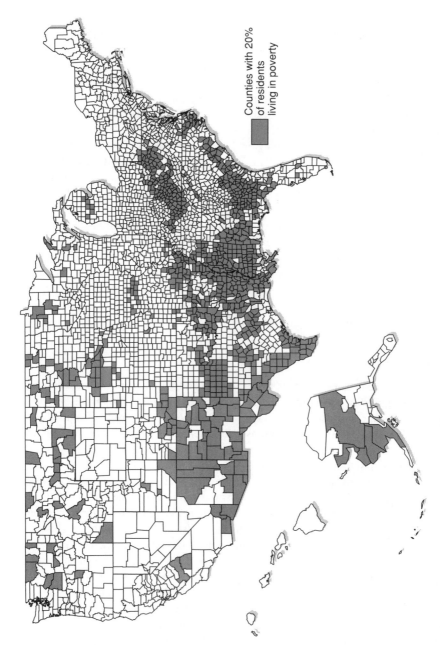

Figure 11.2 The geography of poverty in the United States, 1990

Counties with 20% of residents living in poverty

the lower Mississippi river valley, the Rio Grande river valley, and central Alaska stand out as a rural landscape of pervasive poverty. Scattered counties, often correlated with Native American reservations in the North-West and upper Midwest, are also home to large numbers of rural poor.

The visual presence of rural poverty is far less intensive than in urban contexts. In areas with large numbers of poor people, pockets of ill-maintained houses may be found on backroads or in isolated areas, but the low density rural landscape may mean that physical evidence of the poor is masked by the surroundings. Jackle and Wilson (1992) observe that derelict farmsteads and settlements, homes to the rural poor, are widely romanticized by urbanites as 'quaint' and a nostalgic element of the bucolic rural landscape. But rundown apartments or houses in the inner city are likely to be perceived as threatening places.

In a similar fashion, the economic and social contexts of the urban and rural poor are markedly different. The urban poor are generally unemployed or lack the skills to obtain full-time work. Among the rural poor, the percentage of working poor is very high. The working poor are defined as persons employed full-time or seasonally whose incomes are below official poverty levels. For instance, nearly one-fifth of poor rural householders in the United States work full-time, year-round. The number of disabled or ill adults who were physically unable to work was also much higher among the rural poor than urban poor. Finally, a much smaller number of the rural poor live in female-headed households.

The global economic restructuring beginning in the late 1970s has had an especially damaging impact on the rural working poor. The expanding international competition in commodity production and manufacturing affected rural areas in North America disproportionately. This is particularly true for rural communities without diversified economic bases. Those rural communities overly dependent on one or two economic activities or individual firms were especially vulnerable to structural change. Although farm economies have experienced some negative impacts, most of the more serious and longer-term impacts are focused on non-farm sectors. As manufacturing jobs were shifted to lower wage locations and commodity production jobs were reduced by mechanization, rural unemployment increased dramatically (Wimberly, 1993).

The recent economic downturn is generally considered to be structural rather than cyclical, with the greatest impacts on younger workers, low skill workers, minorities and women. In urban areas the growth in high wage jobs in the service sector has mitigated overall job losses, but rural areas have not experienced this job growth. The result has been an increase in the working poor, up 20% in the United States over the past decade.

Government policy initiatives designed to assist rural areas have largely been ineffective in meeting the needs of the rural poor. One widely held explanation for the failure is the dominant approach used by Western governments. The approach is straightforward: rural development policy is agricultural policy. Subsequently, although there is a critical need for programmes and policies for emphasizing human resource development and social revitalization, the bulk of rural finance goes to farm households. In the United States, the family-directed aid is focused upon the farmer growing a narrow range of agricultural commodities. Non-farm residents and farmers producing other crops are deprived of potential benefits (Wimberly, 1993). Agricultural-based rural development policies have only limited value for the rural poor and disadvantaged, and may weaken the opportunity for long-term economic growth (Vogeler, 1981) and reinforce existing social relations that disempower the poor and disadvantaged.

RURAL HEALTH SERVICES

One area where the disparity in the provision of services between urban and rural areas is well documented is health care. Rural areas present a sharp contrast to urban and suburban jurisdictions with regard to health status and the provision of health care resources. The inequalities between rural and urban areas are not entirely straightforward. Instead there is an intrarural health care gradient. Wealthier rural locales and urban fringe areas have a wider array of health-related services, whereas poorer communities and geographically isolated areas have the lowest levels of health care delivery.

In macro terms, however, a rural–urban dichotomy in health resources is persistent and in some instances is getting more disparate. In the United States, for instance, 22.5% of the population lives in non-metropolitan areas, but only 13.2% of patient care physicians and 6.7% of hospital-based physicians are practising in these areas (Ricketts *et al.*, 1994). Reversing growth trends from the 1970s, rural hospitals and clinics across the United States began to close at an increasing rate during the latter half of the 1980s (Mullner and McNeil, 1986). These closures, in turn, have restricted access or lowered the level of service available to rural Americans, especially when compared with health care services for urban residents. Over a longer period, facility cutbacks further erode the attractiveness of a place in maintaining or recruiting new health care professionals.

Residents of rural areas report lower health status than urban and suburban dwellers. With large numbers of the poor and elderly, rural residents have higher rates of chronic diseases. Accident rates are also higher in the countryside. The combination of large at-risk populations in rural areas and a maldistribution of health care human resources and facilities further stresses the health of rural residents. The recent spread of HIV and tuberculosis in portions of the rural United States lends further significance to questions surrounding the public health needs of rural locales facing expensive, contagious disease threats.

Although universal health care is a fundamental part of the social service system of many industrialized nations, access to health care is not necessarily equally available to rural and urban residents. Lower population densities and distance barriers often mean limitations in quality health care for large segments of rural populations. In Canada, for example, the number of hospital beds per capita in rural and remote rural areas was on a par with the rest of the nation, with the number of physicians and surgeons per capita inversely related to remoteness (Government of Canada, 1995). In 1991 there were three times more physicians per 1000 people in Canadian cities than in the northern hinterlands of the provinces and territories. There was a relative shortage of physicians in the Yukon and the Northwest Territories.

Similarly, Australia government health care planning decisions, based upon per capita formulae, tend to ignore higher than average rates of chronic disease in rural areas and spatial isolation of Australian rural life. Consequently, 'the chronically ill, the mentally ill and disabled persons . . . have neither access to the same services nor to the same level of services currently available to those residing in metropolitan areas' (Lawrence, 1990, p. 116). The Australian Auditor General has reported gross inequalities and inequities in health care provision throughout rural Australia. The most severe disadvantages in services are experienced in the smallest towns.

In a pattern repeated throughout the industrial world, the life expectancy for Native Australians (Aboriginals) is much shorter than the life expectancy for White Australians. Non-Aboriginals live some 15 to 20 years longer. This difference is directly attributable to the lower level of appropriate health services for Aboriginal communities in rural Australia (Lawrence, 1990).

Physician distribution

Although the supply of health care providers (e.g. physicians) is not a perfect measure for determining the quality or provision of health care, it does represent a fundamental element to any health care delivery system. Physician distribution studies do not present a positive perspective for rural areas. For example, longitudinal data indicate a diminishing physician presence in rural America. Using a rural–urban continuum based upon a US Department of Agriculture coding format, Frenzen (1991) analysed physician to population data for 1975, 1981 and 1988. The analysis indicated that physician supply during the three periods was quite similar, with a decline in the availability of medical doctors as rurality increased. Rural non-metropolitan areas were the most underserved places in the United States during the three periods. Although using a different rural–urban classification format, Fruen and Cantrell (1982) reached similar conclusions in parallel research that tracked the supply of primary care physicians between 1950 and 1978. Their research indicated that non-metropolitan areas with the smallest populations consistently experienced the highest population per physician ratios whereas the largest metropolitan areas consistently had the lowest.

The physician location disparity is partially accounted for by extending earlier geographical differences to the present. Thus, current inadequacies reflect past spatial patterns. More troubling, however, are the current trends in health care provider location, combined with the closure of rural medical facilities, which raises the spectre of further rationalization of rural health services in selected rural areas. Studies examining the locational decisions of physicians show that decision making is strongly influenced by the availability of medical infrastructure (Fig. 11.3, overleaf). New physicians in rural areas tend to settle in communities with an existing health care system. Especially important is the presence of other physicians. A survey

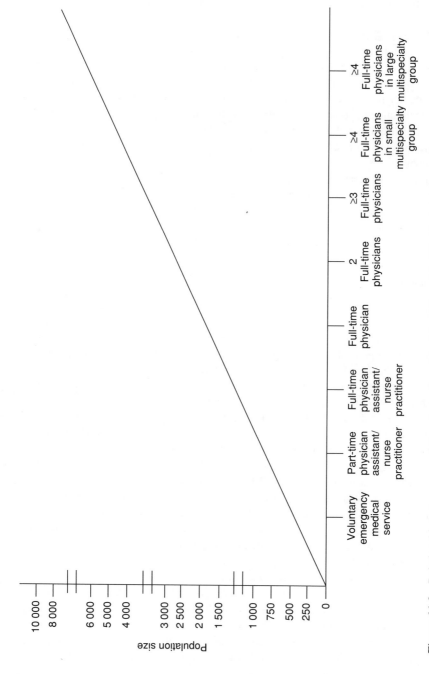

Figure 11.3 Relationship between population size and rural health services (Modified from Baldwin and Rowley, 1990)

by Madison and Combs (1981) of 900 young physicians practising in small US towns found that 61% of the physicians were inclined to establish a practice in a community with four or more other physicians, whereas only 7% would locate in a community with one other physician. Without some form of financial subsidy, only a small proportion of the surveyed physicians would consider moving to the smallest communities.

Physician outmigration from rural areas has also been identified as a serious problem in Australia. Overwork and isolation have been identified as causal factors for the shortage of physicians in small rural towns. In New South Wales (NSW) the ratio of physicians to population in metropolitan areas is 1:30, whereas in western NSW it is 1:1500. One-half of the rural hospitals in NSW lack basic diagnostic equipment and the staff to treat respiratory and heart diseases (Lawrence, 1990).

The policy indications for health care delivery systems are significant. Small rural communities or towns which have lost medical infrastructure are handicapped in their efforts to maintain adequate health care systems. Without government intervention it may be impossible to maintain an effective set of services. For many of the smallest rural communities, the substitution of physician assistants and certified nurses for physicians has already been accomplished. This shift to a lower level of medical service reflects a restructuring of health services. In the face of a declining supply of professional and financial resources, this is a choice that other larger rural communities are increasingly approaching.

RURAL HOUSING

Although the provision of adequate shelter may be seen by some observers as a fundamental right, the availability of affordable and healthy housing is an elusive goal for some rural residents. Rural housing was traditionally synonymous with farm housing, and the problems of rural housing were limited to issues relating to the larger agricultural economy (Rogers, 1983). But with the declining significance of agricultural employment and the growing number of non-traditional rural residents, housing problems have taken on an entirely new meaning. The nature and magnitude of housing-related problems are highly variable between regions and nations. The scope and variety of housing issues are outlined in the following discussions examining rural housing concerns in Canada, the United States and Britain.

Canada

Private ownership of rural housing is quite high in Canada. In 1991, 80% of all rural households owned their own homes. Rural home ownership is 20% above urban home ownership. A comparison of rural housing conditions in Canada between 1981 and 1991 found that housing quality had improved during the period. Housing conditions were evaluated using three core housing criteria: a suitability norm, indexing the amount of sleeping space per family; an adequacy norm, relating to repair needs and the availability of hot and cold running water, bathing facilities and indoor toilets; and an affordability norm, based upon a maximum of 30% of income expenditure for housing. Using these standards, Canadian researchers found rural households with housing needs constituted a 14% share of the national housing need in 1991 (Government of Canada, 1995). This was an improvement from 1981, dropping from 18%.

The most significant rural housing need in Canada was affordability. This was a problem for over 60% of rural households. Housing inadequacy was also a problem for slightly more than 20% of rural households. Rural housing problems were most serious for native populations (Indian, Inuit and Metis). Canada's most rural population (46% of native peoples live outside urban areas) had the highest levels of inadequate housing.

United States

Housing quality in the rural United States has improved dramatically over the past 20 years, but remains a concern in many rural areas. In 1970, over 3 million rural housing units were judged to be substandard; by 1990 the number was down to approximately 1 million dwellings. In the United States, substandard housing is conventionally defined as housing either lacking complete plumbing or overcrowded (i.e. more than 1.1 persons per room). Unfortunately, this standard does not consider the structural characteristics of housing, which may have deteriorated or be dilapidated. If structural inadequacies were evaluated, the unsatisfactory rate would be substantially higher (Housing Assistance Council, 1994, p. 29).

Despite these improvements, housing quality remains a concern for many rural Americans (Table 11.1). Nationally 4.8% of rural dwellings are crowded and/or lacking plumbing. Slightly more urban dwellings fail the same quality standards, 5.5%. Homes with multiple housing quality problems are disproportionately located in rural America; almost half the housing units with crowding and plumbing problems are rural homes.

The problem of poor rural housing conditions is most severe in Alaska, Arizona and New Mexico. Although these three states contain only 1.5% of the rural housing in the United States, they encompassed 27% of substandard rural housing in 1990. Undoubtedly, the magnitude of poor housing in these three states is related

Table 11.1 Rural and urban housing quality indicators in the United States

Indicator	Percent of total rural dwellings	Percent of total urban dwellings
Overcrowded	2.9	5.1
Incomplete kitchen	2.1	0.7
Lack complete plumbing	2.7	0.5

Source: Adapted from Housing Assistance Council (1994).

to the higher than average rates of rural poverty in all three areas. Arizona and New Mexico have the United State's highest proportion of rural poor, 27%. Alaska's 12% is lower, but still higher than the US average. Moreover, beyond the extreme poverty, the remoteness of rural areas in each state presents additional barriers to creating inexpensive, satisfactory shelter.

The cost of housing and the availability of cheap housing are often overlooked when considering the geography of the poor. Fitchen (1991) has documented the prominent role that housing opportunity plays in the migration and lifestyle choices of rural Americans. Low cost housing attracts poor people to rural America and discourages outmigration. Poverty attracts poverty (Jackle and Wilson, 1992). The cheapest housing costs are in rural areas meeting one or more of the following criteria: (1) locales suffering from economic downturns, resulting in a lowering of the housing demand; (2) communities without land development codes or housing quality standards, allowing substandard housing to be marketed; (3) geographically isolated areas, with weaker, lower value land markets; and (4) areas with larger numbers of poor people where it is possible to share accommodation and reduce living costs. In the latter case, the settlement impacts are magnified; that is to say, the large concentrations of low cost housing discourage settlement by potential wealthier residents, thus depressing housing demand further and lowering prices.

Home ownership in the rural United States has been slowly growing for several decades. In 1990 the ownership rate in rural areas was 81%, significantly higher than the 59.1% for urban areas. The 1990 home ownership data shows little change from 1980. Among rural Americans, home ownership is highest in households living in remote areas. Rural village and town residents are less likely to own their dwelling. One cause for the higher proportion of home ownership in rural America, as compared to cities, is the mobile home. Mobile homes or trailers are the fastest growing type of housing unit on the American landscape. In 1990, 4.7 million mobile homes provided shelter to 16.5% of all rural residents. The number of mobile homes grew by 61% over the previous decade, whereas the rural population expanded by only 3.6%.

The expansion of mobile home living is linked to two trends in rural America. The most important is related to economics: mobile homes are a cheaper form of housing than either traditionally constructed dwellings or renting a home. With costs of one-half to one-third as much as a traditional dwelling, mobile homes represent affordable housing for lower income residents. The price differential is most pronounced in remote rural areas and areas with challenging physical terrain (e.g. steep topography, poor soils) that increases construction-related expenses. A second trend is demographic. Mobile homes are a popular housing option for elderly persons resettling in rural retirement areas. They are seen as providing comfortable living space with lower maintenance expenditures. The accommodation of mobile homes by planning authorities in rural jurisdictions has facilitated the widespread acceptance of this newest housing form.

The spatial pattern of mobile home living in the United States reflects these two powerful forces. The heaviest concentration of mobile home residents is seen in a broad arc of states across the South-East and West (Fig. 11.4, overleaf). These

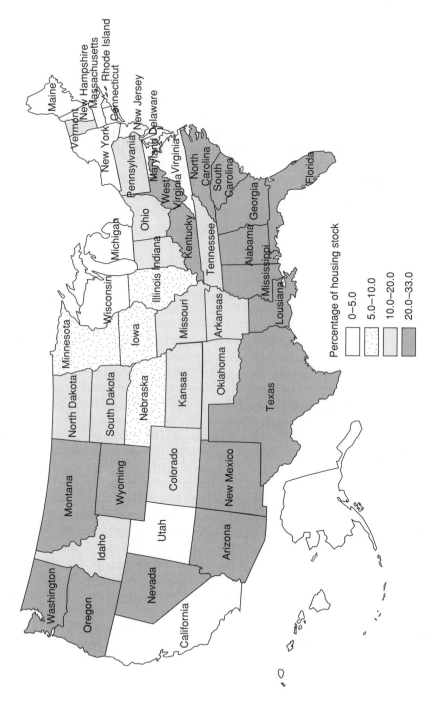

Figure 11.4 Rural US population living in mobile homes (Modified from Housing Assistance Council, 1994)

Percentage of housing stock

- 0–5.0
- 5.0–10.0
- 10.0–20.0
- 20.0–33.0

'sunbelt' jurisdictions hold the largest number of the rural poor as well as the greatest concentration of retirees.

Britain

In Britain the key issues for housing provision are related not to quality but to quantity. A review by Robinson (1992) indicated an escalating problem of housing shortages in rural Britain, especially for the rural poor and working classes. Constraints on rural housing opportunities are a result of complex economic, institutional and social forces that have evolved since the end of World War II, but they are most often linked with urban migration to the countryside (counterurbanization) and increased rural home ownership by newcomers. Counterurbanization was first recognized in the 1960s and has continued through to the present. Today, rural growth is strongest in a semicircular belt extending about 100 to 190 kilometres from London (Sherwood and Lewis, 1993).

The in-migration of middle-class households accompanied by the growing 'second home' market has reduced the housing market options for long-term rural residents. The greatest housing pressures are focused on small towns and villages located in the picturesque countryside, but within commuting distance of urban centres (Marsden *et al.*, 1993). And land-use plans, favoured by new residents, simultaneously restrain the availability of sites for new housing, contributing to higher prices for land and higher housing prices. As a consequence, ownership has become an ever elusive option in rural Britain. Recent data indicates that only about one-third of the households in counties near London can afford to purchase homes. At the same time, the rental housing market has shrunk because of the sale of formerly rental properties and the government's 'right to buy' policy for tenants of council housing (Shucksmith, 1991).

The diminished supply of rental housing and of affordable houses for sale has contributed to larger problems of social cleavage in rural Britain. Most of the attention and political conflict has been focused on providing housing for the counterurbanization market. None of the actors involved in the rural land development arena has shown any particular interest in serving the rental and home ownership needs of lower income rural residents (Marsden *et al.*, 1993).

CONCLUSION

This chapter began with an overview of the recent political and economic changes that have swept across the rural landscape, affecting social relations and service provision. The impact of these changes has been to accelerate social unevenness between urban and rural areas as well as between prosperous and declining rural locales. As the 'have not' rural areas continue to decline and the rural disadvantaged and poor populations grow, the question of what to do becomes more pressing. Creative and effective solutions for solving rural (and inner city) human resource and service provision problems have not been abundant. There is a great risk that the rural poor and their needs will remain invisible, or even worse, they will be ignored like the problems of the inner city. The challenge for policy makers is to construct rural development strategies that meet the human resource needs of rural communities and allow them to build a long-term sustainable future. In the present political and economic climate, this will not be an easy task.

REFERENCES

Baldwin, D.C. and Rowley, B. (1990) Alternative models for the delivery of rural health care: a case study of a western frontier state. *Journal of Rural Health*, **6**, 256–72.

Banfield, E.C. (1958) *The unheavenly city*. Little, Brown & Co, Boston.

Bradley, T. (1986) Poverty and dependency in village England. In Lowe, P., Bradley, T. and Wright, S. (eds) *Deprivation and welfare in rural areas*. GeoBooks, Norwich, pp. 151–74.

Bradshaw, T. (1993) In the shadow of urban growth: bifurcation in rural Californian communities. In Lyson, T. and Falk, W. (eds) *Forgotten places: uneven development in rural America*. University Press of Kansas, Lawrence KS, pp. 218–56.

Bunce, M. and Walker, G. (1992) The transformation of rural life: the case of Toronto's countryside. In Bowler, I., Bryant, C. and Nellis, D. (eds) *Contemporary rural systems in transition*, vol 2, *Economy and society*. CAB International, Wallingford, pp. 49–61.

Cloke, P. (1992) Some initial thoughts on culture and the underclass. In Bowler, I., Bryant, C. and Nellis, D. (eds) *Contemporary rural systems in transition*, vol 2, *Economy and society*, CAB International, Wallingford, pp. 29–48.

Daniels, T. and Lapping, M. (1987) Small town triage: a rural settlement policy for the American Midwest. *Journal of Rural Studies*, **3**(3), 273–80.

Duncan, D. (1992) *Rural poverty in America*. Auburn House, Westport CT.

Eyles, J. (1987) Poverty, deprivation and social planning. In Pacione, M. (ed) *Social geography: progress and prospect*. Croom Helm, Beckenham, pp. 201–251.

Falk, W. and Lyson, T. (1988) *High tech, low tech, no tech: industrial and occupational change in the South*. State University of New York Press, Albany NY.

Fitchen, J. (1991) *Endangered spaces, enduring places: change, identity, and survival in rural America*. Westview, Boulder CO.

Fitchen, J.M. (1981) *Poverty in America: a case study*. Westview, Boulder CO.

Frenzen, P.D. (1991) The increasing supply of physicians in US urban and rural areas, 1975 to 1988. *American Journal of Public Health*, **81**, 1141–47.

Fruen, M.A. and Cantrell, J.R. (1982) Geographic distribution of physicians: past trends and future influences. *Inquiry*, **19**(1), 44–50.

Garfinkel, I. and McClanahan, S. (1994) Single-mother families, economic insecurity and government policy. In Danziger, S.H., Sandefur, G.D. and Weinberg, D.H. (eds) *Confronting poverty, prescriptions for change*. Russell Sage Foundation, New York, pp. 205–25.

Government of Canada (1995) *Rural Canada: a profile*. Government of Canada, Ottawa.

Halfacree, K.H. (1993) Locality and social representation: space, discourse and alternative definitions of the rural. *Journal of Rural Studies*, **9**(1), 23–37.

Harrington, M. (1962) *The other America: poverty in the United States*. Macmillan, New York.

Hirschman, A. (1970) *Exit, voice, and loyalty*. Harvard University Press, Cambridge MA.

Housing Assistance Council (1994) *Taking stock of rural poverty and housing for the 1990s*, Housing Assistance Council, Washington DC.

Hudson, R. and Sadler, D. (1986) Contesting works closures in western Europe's old industrial regions: defending place or betraying class? In Scott, A. and Storper, M. (eds) *Production, work, and territory: the geographical anatomy of capitalism.* Allen and Unwin, Boston, pp. 265–80.

Jackle, J.A. and Wilson, D. (1992) *Derelict landscapes, the wasting of America's built environment.* Rowman and Littlefield, Savage MD.

Kelso, W. (1994) *Poverty and the underclass: changing perceptions of the poor in America.* New York University Press, New York.

Lawrence, G. (1990) Agricultural restructuring and rural social change in Australia. In Marsden, T., Lowe, P. and Whatmore, S. (eds) *Rural restructuring, global processes and their responses.* David Fulton, London, pp. 101–28.

Lewis, G.J. (1983) Rural communities. In Pacione, M. (ed) *Progress in rural geography.* Croom Helm, Beckenham, pp. 149–72.

Lewis, O. (1966) The culture of poverty. *Scientific American,* **215**, 19–25.

Lonsdale, R. and Enyedi, G. (1984) *Rural public services: international comparisons,* Westview, Boulder CO.

Lowe, P., Bradley, T. and Wright, S. (eds) (1986) *Deprivation and welfare in rural areas.* GeoBooks, Norwich.

McLaughlin, B. (1986) The rural deprivation debate: retrospect and prospect. In Lowe, P., Bradley, T. and Wright, S. (eds) *Deprivation and welfare in rural areas.* Geobooks, Norwich, pp. 43–54.

Madison, D.L. and Combs, C.D. (1981) Location patterns of recent physician settlers in rural America. *Journal of Community Health,* **6**(4), 267–74.

Marsden, T., Murdoch, J., Lowe, P., Munton, R. and Flynn, A. (1993) *Constructing the countryside.* Westview, Boulder CO.

Mincy, R.B. (1994) The underclass: concept, controversy, and evidence. In Danziger, S.H., Sandefur, G.D. and Weinberg, D.H. (eds) *Confronting poverty, prescriptions for change.* Russell Sage Foundation, New York, pp. 109–46.

Moseley, M. (1979) *Accessibility: the rural challenge.* Methuen, London.

Moseley, M. (1980) Is deprivation really rural? *The Planner,* **66**, 97.

Mullner, R.M. and McNeil, D. (1986) Rural and urban hospital closures: a comparison. *Health Affairs,* **3**(5), 131–41.

Myrdal, G. (1963) *The challenge of affluence.* Pantheon, New York.

Organization for Economic Cooperation and Development (1991) *New ways of managing services in rural areas.* OECD, Paris.

Pacione, M. (1984) *Rural geography.* Croom Helm, London.

Pacione, M. (1995) The geography of deprivation in rural Scotland. *Transactions of the Institute of British Geographers,* **20**(2), 173–92.

Pawson, E. and Scott, G. (1992) The regional consequences of economic restructuring: the West Coast, New Zealand 1984–1991. *Journal of Rural Studies,* **8**(4), 373–87.

Peck, J. and Tickell, A. (1994) Jungle law breaks out: neoliberalism and global–local disorder. *Area,* **26**(4), 317–26.

Popper, D.E. and Popper, F.J. (1987) The Great Plains: from dust to dust. *Planning,* **53**, 13–18.

President's National Advisory Commission on Rural Poverty (1967) *The people left behind.* US Government Printing Office, Washington DC.

Ricketts, T., Savitz, L., Gesler, W. and Osborne, D. (1994) *Geographic methods of health services research: a focus on the rural–urban continuum.* University Press of America, Lanham MD.

Robinson, G. (1992) The provision of rural housing: policies in the United Kingdom. In Bowler, I., Bryant, C. and Nellis, D. (eds) *Contemporary rural systems in transition*, vol 2, *Economy and society*. CAB International, Wallingford, pp. 110–28.

Rogers, A.W. (1983) Housing. In Pacione, M. (ed) *Progress in rural geography*. Croom Helm, Beckenham, pp. 106–29.

Sandefur, G. (1989) American Indian reservations: the first underclass areas? *Focus*, **12**, 37–41.

Sengenberger, W. (1993) Local development and international competition. *International Labour Review*, **132**, 313–29.

Shaw, J.M. (1979) Rural deprivation and social planning. In Shaw, J.M. (ed) *Rural deprivation and planning*. GeoBooks, Norwich.

Sherwood, K. and Lewis, G. (1993) Counterurbanization and the planning of British rural settlements: the integration of local housing needs. In Harper, S. (ed) *The greening of rural policy: international perspectives*. Belhaven, London, pp. 167–84.

Shucksmith, M. (1991) Still no homes for locals? Affordable housing and planning controls in rural areas. In Champion, A.G. and Watkins, C. (eds) *People in the countryside: studies of social change in rural Britain*. Chapman, London.

Vogeler, I. (1981) *The myth of the family farm: agribusiness dominance of US agriculture.* Westview, Boulder CO.

Wimberly, R.C. (1993) Policy perspectives on social, agricultural, and rural sustainability. *Rural Sociology*, **58**, 1–29.

CHAPTER 12

CONCLUSION

Brian Ilbery

This book has attempted to examine some of the processes and outcomes of recent change experienced by rural areas in developed market economies. Little attempt has been made to enter into the debate on what is rural; instead, the emphasis has been on exploring the geography of rural restructuring and the trend towards a post-productivist countryside. The objectives of this short concluding chapter are first to summarize the nature of rural change and the main processes at work, secondly to highlight the possible tensions and conflicts which these changes may have engendered, and thirdly to outline the prospects for sustainable rural development in the future.

Drawing together the different chapters of the book illustrates the dynamic nature of economic and social change in the rural areas of many developed market economies. Some of the main changes are as follows:

- The declining significance of agriculture in terms of employment and the relative importance of food production.
- The growing importance of pluriactivity and quality food products as part of the post-productivist transition in agriculture.
- The continued afforestation of agricultural land and the changing values and perceptions of forestry as a land use.
- The increasing significance of employment in manufacturing, high technology and service industries, especially in the form of SMEs.
- The emergence of new uses of rural space, including retailing, tourism, recreation and environmental conservation.
- The repopulation of rural areas, especially by the service classes, but also the continued outmigration by young people.
- The increasing differentiation in the quality of life between the 'haves' and 'have nots' in rural areas.

Not all rural areas have been affected to the same extent by these trends; indeed, they will respond in different ways to the processes affecting society in general according to the diversity of their local rural conditions. The major processes of change, therefore, are both 'external' and 'internal' to rural areas. It is clear that rural areas are components of much broader socio-economic transformations, including the globalization of capital restructuring, geopolitical reorganization, international residential mobility and the internationalization of trade. International competition is increasing and the ideology of the free market is leading to a search for new forms of governance, which are less dependent on direct state intervention and more reliant upon popular participation and private entrepreneurship. Thus state deregulation and privatization are increasing, although there is some reregulation in areas of environmental concern and sustainable development.

Environmentalism is a rising political force. Together with new technologies (e.g. biotechnology and information and communication technologies) and other processes of change, environmentalism is transforming rural space and creating various opportunities and constraints for future development. For example, many parts of the countryside have become sanitized and commodified; a particular image is being developed and packaged, often by 'external' entrepreneurs, and sold to tourists. Rural identity emerges from power relationships, especially at the national level, and specific groups are utilizing imagery to project messages which favour themselves. There is little doubt that long-held associations made about rural areas are now under challenge; this in turn can bring anxiety to many people living in the countryside. Nevertheless, it needs to be stressed that rural space is made by groups of both local and non-local people; thus local and 'internal' factors cannot be ignored. Indeed, the emphasis within rural policy is very much towards increasing localization, but within the macroprocesses transforming society.

The cultural values of rural areas are also changing. New cultural identities are being placed on rural space by social groups (especially the service classes) moving into rural areas. A new rural idyll is being created by the wealthy for the wealthy, and the service classes are having an increasing influence over the physical and social character of rural areas. For example, they often see the countryside as a zone of consumption available for recreation rather than as an area of primary production. Some parts of rural space have become gentrified and certain groups (notably the original population and rural 'others') are being excluded and marginalized. Processes of selectivity and exclusion are very real in rural areas.

As a consequence, rural areas have in many ways become arenas of conflict and tension. The countryside is being contested and rural change has promoted a number of dualities:

- consumption versus production
- development versus preservation
- deregulation versus reregulation
- global versus local processes
- the service classes versus the 'locals'
- the farm lobby versus environmentalists

Such dualities have encouraged new economic tendencies and cultural forms; as a result, rural space has become more strongly differentiated. Rural areas display different capacities for accommodating change and thus embark on different development trajectories. Marsden (Chapter 2) conceptualized such development trajectories into four ideal types; of these, it is clearly in the 'contested' countryside where many of these dualities are enacted.

As rural areas respond to economic and social change and become more consumptive than productive, policies for rural development need to be reassessed. A series of influential reports concerning future directions in rural planning and policy have been published during the 1990s and there is a search for new forms of governance which are of relevance to rural areas. In the past, policy makers' ideas of 'rural problems' often appeared not to be shared by the rural residents; the approach was 'top down' rather than 'bottom up'. In the future, policies for rural development need to share three characteristic features: they need to be multisectoral; they must be sustainable; and they should be based on the principle of subsidiarity.

Agriculture is of declining relative importance in rural areas and is now competing with other economic activities, in terms of employment and income generation. Consequently, the rural economy can no longer be regulated by sector-specific institutions and policies. Policy making requires an integrated approach that crosses sectoral divides. Agriculture needs to be just one element of such a multisectoral development policy for rural areas. This element would promote quality products and farm diversification, as well as transferring support to farmers away from economic subsidies and towards environmental schemes. However, it is vital to involve other sectors in rural policy in order to create jobs, a better quality of life and prosperity.

Environmental and social conditions would also need to be integrated into such economic decision making. Indeed, a sustainable policy for rural development would take account of all aspects of rural life – economic, social, cultural and environmental. Sustainability in this sense means maintaining economic production and enhancing socio-economic wellbeing, without causing damage to local ecosystems. Change rather than fossilization is an important component of sustainable rural communities. This involves integrating newcomers into the community, reassessing the role of established institutions such as the Church and the Women's Institute, and monitoring the effects of deregulation and new forms of governance on people in rural areas.

The focus of integrated rural development policies should be on local people. How can local people exploit rural resources in a sustainable fashion to allow wealth generation at the local level? Local determination is central and therefore so is the principle of subsidiarity. Thus, rural development strategies need to be geared to local circumstances and to emphasize local initiative and enterprise. However, as well as retaining local autonomy, rural areas need to seek integration into the global economy. This is not easy and rural areas will need to create products which depend on local identity (geographical origin) for their market niche, but which can be exported beyond the local region. One example would be the production of high quality food products with clear traceability, which would 'fit in' with the increasing trend towards careful consumption following health scares over BSE and *E. coli*.

Finally, and related to the local dimension of rural development, policies need to be geared to different types of area. The future for remote, mountainous areas and those still heavily dependent on agriculture (with few alternatives) is far from clear. This is a clear reason why future development policies must be localized, so that local people are empowered to define their own needs and share in the development of the services required to meet them.

It is quite easy to state that any future rural development policies should be multisectoral, sustainable and localized. However, it is much more difficult to actually develop such policies and to build a long-term sustainable future for rural areas. Hopefully, this book has presented some of the main processes and outcomes of change in rural areas, ideas which can act as guidelines for policy development.

INDEX